FUNDAMENTALS OF ANIMAL BEHAVIOUR

FUNDAMENTALS OF
ANIMAL BEHAVIOUR

K. S. S. MOOSATH
AJIN S
RENJITH THOMAS

KNOWLEDGE
BAKERS

KNOWLEDGE BAKERS

ISBN: 978-93-49964-92-1

Published by : **Knowledge Bakers**
Pune, India
Email: knowledgebakers2019@gmail.com

Preface

Understanding animal behavior is fundamental to advancing fields such as animal husbandry, wildlife conservation, veterinary science, and animal welfare. Animal behavior influences how animals interact with each other, their environment, and humans. Whether in the context of domesticated animals, wild species, or laboratory settings, the study of animal behavior provides key insights into the biological and psychological mechanisms that drive actions, reactions, and adaptations. This knowledge is critical for improving animal welfare, enhancing agricultural practices, managing wildlife populations, and fostering human-animal interactions.

This book, titled Fundamentals of Animal Behaviour, aims to provide a comprehensive introduction to the principles, theories, and applications of animal behavior. Designed for students, researchers, and professionals in fields like biology, psychology, veterinary science, and animal management, this text serves as both a foundational resource and a practical guide to understanding how animals behave and why.

The book begins by exploring the foundational concepts of animal behavior, including the evolution of behavior, genetic and environmental influences, and the neurobiological mechanisms that underpin various behaviors. It introduces the key approaches to studying animal behavior, including observational, experimental, and ethological methods, with a focus on real-world applications.

Fundamentals of Animal Behaviour covers a wide range of behavioral topics, such as communication, social behavior, mating systems, foraging, territoriality, and predator-prey interactions. It also explores the cognitive and emotional aspects of behavior, examining how animals perceive and process information, learn, and adapt to their environments. The role of instinct versus learning in shaping behavior is also a central theme, helping readers understand the complexities of behavioral responses.

A significant portion of the book is dedicated to the application of animal behavior in practical contexts. It addresses the importance of understanding animal behavior in agriculture, particularly in managing farm animals, improving livestock production systems, and enhancing animal welfare. The book also explores the behavioral needs of companion animals, zoo animals, and wildlife, highlighting the importance of meeting these needs in captivity and ensuring that animals lead fulfilling lives. Ethical considerations surrounding the study and management of animal behavior are also discussed.

Through a combination of theoretical insights, real-world examples, and case studies, Fundamentals of Animal Behaviour provides readers with the tools needed to analyze and interpret animal behavior in both natural and controlled settings. By focusing on the relationship between behavior, environment, and welfare, the book offers a holistic approach to understanding animals and their actions.

It is our hope that this book will serve as a valuable resource for anyone interested in animal behavior, offering a solid grounding in the subject while encouraging further exploration into the fascinating and diverse world of animal actions, interactions, and emotions.

K. S. S. Moosath
Ajin S
Renjith Thomas

Contents

1

Introduction

This chapter introduces the main topics that are covered by this field of research. A historical account discusses how we have tried to explore the behavior of humans and animals, to understand the biological constraints that impact our cognitive potential, and to make sense of the nature of human and animal personalities. As a scientific discipline Animal Behavior conforms to the rigors of the Scientific Method.

DEFINITION OF ANIMAL BEHAVIO]R

Animal Behavior is the scientific study of the wild and wonderful ways in which animals interact with each other, with other living beings, and with the environment. It explores how animals relate to their physical environment as well as to other organisms, and includes topics such as how animals find and defend resources, avoid predators, choose mates, reproduce, and care for their young.

Previous lesser definitions of Animal Behavior include:

"Behavior is motion". "Movement, not necessarily movement of the whole animal... muscular contractions" or "The whole function of the nervous system can be summed up in one word, conduction." - In Sherington's view the basic element of behavior was formed in a reflex-arc, where receptor organs receive sensory stimuli and are conducted to an effector organ. This highly reductionist position has received criticism from many angles, specifically, from arguments that it fails to account for spontaneous behaviors, behaviours that are characterized by a lack of motion, as well as disregarding the complexity of emergent properties in behavior.

"What a plant or animal does, in the course of an individual's lifetime, in response to some event or change in its environment." - This definition reduces behavior to phenotypic plasticity, and is thus not specific enough.

"Behavior is all observable or otherwise measurable muscular and secretory responses (or lack thereof) and related phenomena in response to changes in an animal's internal or external environment." Behavior may include components which do not lend themselves to simple quantification.

"Behavior is characterized by entropic and energetic transductions by an organism, in which the long-term averages convert high entropic and low energetic sensory inputs into low entropic and high energetic outputs." - This definition is too incomprehensible to be of great use.

Scope of Animal Behavior

The study of the behavior of animals is very fortunate that students, who enter this field with a wish to learn more about its scientific approaches, are already intrigued by their own observations. Invariably, beginning students come in equipped with detailed descriptions of behavioral eccentricities of their own pets, wildlife, or smaller siblings. The main challenge is for us to understand the fundamental nature of the scientific approach that must be used to study behavior more formally. Only a rigorous experimental characterization of behavioral phenomena will yield common generalities on how and why animals do what they do - rather than providing a simple catalog of anecdotes. The study of behavior thus critically depends on our ability to phrase our general inquiry as a series of questions that can be answered with standard scientific approaches, that are amenable to experimental probes, and which lead to an explicit and unambiguous interpretation of its results.

Niko Tinbergen's legacy as a exceptionally gifted experimentalist include the phrasing of four different types of questions that can be asked of any animal behavior, now simply referred to as Tinbergen's 4 Aims. This paper summarizes the multidisciplinary approach that had guided the emergence of the field of Ethology since its early roots at the beginning of the 20th century. They group broadly into questions about proximate and ultimate explanations. Proximate explanations are closest to the event that is to be explained, while higher, ultimate explanations instead focus on the deeper reasons for why something happened.

Proximate Explanations

Neural, Endocrine, and Genetic Mechanisms

Causes that explain how a behavior is elicited and coordinated. What are the anatomical or physiological mechanisms that underlie the behavior? The behavior is present because the nervous system makes it happen. It includes a search for the impact of genetic components, physiological mechanisms, environmental conditions needed, and hormones on behavior.

Examples:

Ontogeny, Development: How does the behavior develop and change during the lifespan of a single individual? The behavior is present because the individual developed an ability to do it this way.

Ultimate Explanations

Survival Value

Explanations are phrased as ultimate causes that explain why a behavior has evolved. Selection ought to favor individuals with behaviors that help it increase its reproductive success. The behavior is present because the individual inherited them from parents who were historically successful with this behavior.. This includes a search for functional significance or adaptiveness of behavior.

Phylogeny, Evolution

What were the macroevolutionary patterns of changes that lead to the behavior today?

Phylogeny is the study of evolutionary relatedness among groups of organisms as It treats a species as a group of lineage-connected individuals over time. It attempts to historically characterize the branching processes in which breeding populations have changed over time. The ethologist's view holds that behavioral characteristics, which receive at least a partial genetic contribution, and can be viewed in a phylogenetic perspective like any other morphological trait. The behavior is present because the individual is an offspring from ancestors that had the behavior.

Example - Territoriality in Red-winged Blackbirds

Male red-winged blackbirds are conspicuous and familiar birds throughout most of North America, where they commonly occupy and defend perches near sources of water. Aggressive while defending a territory, they will attack much larger animals, such as ravens, hawks, or humans who encroach upon their territories. In the northern US, male red-winged blackbirds return from their wintering grounds in early March. In rather stunning song-spread display they flaunt their prominent red epaulets, they lower and spread their tails, puff out their feathers, issuing their distinctive Cong-a-lee! call. According to Tinbergen's 4 Aims our interests as to "Why" they would show such a distinctive type of territorial advertising can be organized in a number of different ways.

Aim 1: Proximate Mechanisms

We may ask questions on the particular mechanics that cause the behavior via its underlying neural, endocrine, and genetic mechanisms - i.e., the behavior's proximate causation. Our interests would focus on whether the male is responding to a particular stimulus such as the approach of a rival, what neural mechanisms aid in the detection of key stimuli, what neural circuits, hormonal triggers, or gene products are necessary for the behavior's expression, or how seasonal changes in day length are able to control them?

The vocal organ of birds, the syrinx, is located at the base of the blackbird's trachea. Sound is produced when air flowing through it causes the syringeal walls to vibrate. Muscles modulate the sound shape by changing the tension of the membranes and the bronchial openings. Centers for song production pathways include the High vocal center (HVC); robust nucleus of archistriatum (RA); and the tracheosyringeal part of the hypoglossal nucleus (nXIIts). Both pathways show sexual dimorphism, with the male producing song most of the time. Androgens are believed to play a key role in the expression of male song behaviors as it exhibits a strong correlation with seasonal testosterone profiles. Song is enhanced with testosterone injections and ceases following adult castration. Injections of testosterone induce growth of the HVC and thereby even produce song in female birds, which otherwise would not sing.

Aim 2: Development, Learning

We could also query in what way the territorial display emerges in development. What role does maturation play? To what degree does the behavior depend on learning processes and tutors? What neural substrates are involved in its acquisition?

The song of red-winged blackbird males, like that of many other song birds, is learned. A comparison of hatchlings, which were either deprived of normal song altogether or simply presented with its recording, demonstrated that individuals readily learned their song from conspecifics in the vicinity. Far from a pure cultural transmission of song dialects, however, innate predispositions played an important role. Some individuals were able to learn complex repertoires with minimal stimulation while others were quite reluctant to develop elaborate, species-specific calls.

Some remain faithful to their tutors while others improvised readily and invented new versions. The neural signals that subserve vocal learning center on the anterior forebrain pathway with magnocellular nucleus of the anterior neostriatum (LMAN), Area X, and the Dorsolateral Medial thalamus (DLM).

Aim 3: Functional Significance

Alternatively, we may be interested to know what function this particular territorial display serves. If selection favors individuals with a behavior that is beneficial to the bearer's reproductive success, what are the precise benefits bestowed to that individual and its parents? What defines this behavior as successful in evolutionary terms? What factors are turned up by a search for the functional significance or adaptiveness of this behavior? When red-winged blackbirds were surgically muted, the loss of song affected their ability to defend the territory. Males were captured and replaced by an auditory display of the individual's recorded song to examine whether the broadcasts alone would be sufficient to defend the territories.

Aim 4: Evolution

Does the red-winged blackbird perform its territorial display because its ancestors did it in a similar way? Is it the only one within its clade that has this display? What were the precise patterns of change throughout the evolutionary lineage that result in the behavior today? How does the red-winged blackbird's display compare to that of other blackbirds? Most members of the red-winged blackbird's evolutionary lineage communicate with simple call notes, suggesting that its specific call features may have been inherited from an ancestral taxon.

HISTORY OF ANIMAL BEHAVIOR

Early Influences

Prehistoric Period

Our early ancestors critically depended on a detailed understanding of animal behavior. Specifically, knowledge of habitat preference, movement patterns, and sensory biology of prey as well as predators were essential. Rock paintings, figurines, and carvings, depicting scenes from daily life and religious rituals, commonly feature detailed likenesses of many kinds of animals. Early forms of rock art emerged during the Paleolithic (200,000 BCE), however, the advent of modern man brought about extraordinary cave wall paintings (40,000 BCE) of animals, hunting scenes and other graphic illustrations of prehistoric life.

Holes in Neolithic skulls illustrate that patients dating back to such early times have been subjected to head surgery, presumably with the aim of curing headaches or mental disorders. Sumerian descriptions show the use of Opium, derived from poppy plants, to produce euphoriant effects.

Native American

A common theme contends that the world was populated by many peoples who were subsequently transformed into different kinds of animals. Thus all animals and humans share the presence of a soul. The close bond between them is shown in frequent rituals where the behavior of different animals is simulated. Imagery of animals is common. The Great Serpent Mound represents an effigy mound located on a plateau along Ohio Brush Creek in Adams County, Ohio. Conforming to the topography, the snake's head approaches a cliff point, the snake then winds back and forth for more than 200 meters and ends with a triple coiled tail.

Egyptian Mythology

Ancient Egyptian societies showed a keen interest in the temperaments and sensory abilities of different animals and ascribed them to individual deities. Ancient Egyptians had a highly developed view of the afterlife with elaborate rituals for preparing the body and soul for a peaceful existence after death. Beliefs about the soul and its needs led to numerous ceremonies for the preservation of the body. The brain apparently was not considered an essential element for eternity, as it was treated with a distinct lack of dignity. It was often spooned into small pieces and removed through the nose, then unceremoniously discarded.

Classical Antiquity

Ancient Greek Period

Plato held the notion that the human mind as an entity pre-existed somewhere in the heavens, before being sent down to join a body on earth. Hippocrates believed that the brain was not only involved with sensation but was also the seat of intelligence. Like many before and after him, Aristotle was intrigued by the natural world around him. He, however, stands out as a seminal thinker on a wide range of subjects as he was able to move beyond his substantial abilities at observation by forwarding daring hypotheses in order to categorize and make sense of what he saw. Aristotle provided the first written records of mutualistic ties between individuals, animal tool use, and brood parasitism. His views on human nature considered our behavior as constituting formed habits and a desire for happiness. With his proposed concept of a tabula rasa he forwarded the thesis that individuals are born with no innate or built-in mental content, in a word "blank", and that their entire resource of knowledge is built up gradually from their experiences and sensory perceptions of the outside world. Herophilus studied the nervous system and distinguished between sensory nerves and motor nerves. Erasistratus studied the brain and distinguished between the cerebrum and cerebellum.

Roman Empire

As a physician in a gladiator school, Galen gained a unique view of the behavioral deficits that resulted from acute head trauma. Not surprisingly he wrote extensively on the connection between paralysis and severance of the spinal cord. His many audacious operations, including brain and eye surgery, were not again used for almost two millennia.

Middle Ages

Naturalists and philosophers focused on classical texts and, in particular, on the Bible in Latin. Accepted were Aristotle's views on all matters of science, including psychology. As knowledge of Greek declined, the West found itself cut off from its Greek philosophical and scientific roots. General levels of literacy further decreased as learning became largely restricted to monastic and cathedral schools that were primarily aimed at training clergy. The Church exerted an overpowering role in philosophical thought and nontheological scholarship was rare. Despite this many influences on the study of behavior in later periods began to emerge.

Arabists

Arabian scholars describe cranial and spinal nerves, along with surgical procedures for neurological disorders.

Renaissance

The principal features of the Renaissance (14th-16th century) were the revival of learning based on classical sources and the broad advancements of science. It pioneered scientific study through deductive reasoning. René Descartes supplied a foundation for the pursuit of knowledge.

Unity of all sciences through deductive reasoning

Never accept anything as true that is not known clearly to be such:

Divide problems into as many parts as possible, and proceed from the simplest to the hardest.

He argues that he perceives his body through the use of the senses, which often proven unreliable. So he concludes that thinking is his essence as it is the only thing about him that cannot be doubted. Descartes defines "thought" as every activity of which a person is immediately conscious. In his Mind/body dualism he viewed the body as a machine that follows the laws of physics. In contrast, he regarded the mind (or soul) as a nonmaterial entity that is not subject to such physical constraints. The Cartesians (i.e., the followers of Descartes) aimed to explain behavior using simple, natural, and entirely mechanistic principles. In their view only humans were endowed with an ability to reason.

With the advent of the scientific revolution, medieval worldviews steadily gave way to increasingly analytical methodologies and a weakening of vitalist traditions. Although medical thought continued to be guided by the occult, critical thinking began to chip away at the long-held notions based on Galen. Hippocrates' views are revived and Renaissance artists such as Leonardo da Vinci conducted detailed studies of human anatomy in order to better portray the human form. As Andreas Vesalius corrects Greek medical errors, he revolutionizes medicine. William Harvey applies scientific methodology to human physiology and solves the human circulatory system. Paracelsus rejects occultism and pioneers the use of chemicals and minerals in medicine.

Enlightenment

Thinkers of the Age of Reason (18th century) perceive their efforts as bringing to an end previous periods of irrationality, superstition, and tyranny. John Locke's philosophy revives Aristiotles theory of a tabula rasa, where the (human) mind is at birth a "blank slate" without rules for processing data, and where data is added and rules for processing are formed solely by one's sensory experiences. The concept of a blank mind at birth emphasized the individual's freedom to author one's own human character. Moreover, as members of ones society will become what society make of them, it becomes essential to assure universal access to learning opportunities, medical care, and possibilities for personal development.

Evolution

In the second half of the 19th century the simple taxonomic classification of all life forms was supplanted by a growing focus on parsimonious and scientifically testable (phylegenetic) hypotheses that could explain the relatedness among groups of organisms. Studies of the fossil record, the geographic distribution of faunas, and the characteristics that linked the diversity of live, convinced a growing number of scientists that species indeed changed over time. Evolution became viewed as a branching process, whereby populations are altered over time and may specialize into separate branches, hybridize together, or terminate by extinction. However, the mechanism driving these changes remained unclear until the 1859 publication of Charles Darwin's On the Origin of Species, which provided the theory of evolution with the likely mechanism of natural selection.

Charles Darwin (12 February, 1809 – 19 April, 1882) represents arguably among the most significant scientists of all time. He realized and presented compelling evidence for natural selection as the main mechanism by which all species of life have evolved over time from common ancestors. Evolution, as an explanation for the diversity of life, had emerged earlier and was generally accepted by most in his lifetime. His 1859 book On the Origin of Species established evolutionary descent with modification as the dominant scientific explanation of diversification in nature and this view emerged as the primary explanation in modern evolutionary theory. He also examined human evolution and sexual selection in The Descent of Man, and Selection in Relation to Sex, followed by The Expression of the Emotions in Man and Animals. Darwin suggested that all human characteristics are fundamentally inherited and all have been subject to evolution. Emotions may best be understood as consequences of adaptive reactions.

George John Romanes, a Canadian-born English evolutionary biologist and physiologist, laid the foundation for comparative psychology, postulating a similarity of cognitive processes and mechanisms between humans and animals.

Psychophysics

Psychophysicists, such as Gustav Theodor Fechner and Wilhelm Wundt, aimed to relate matter to mind via a study of inner experiences or feelings towards defined physical stimuli. Psychophysicists explicitly rejected the concepts of vitalists, i.e. that the functions of a living organism are due to a vital principle distinct from physicochemical forces. The main focus lay on the way subjects perceive experimental stimuli that can be objectively measured, such as pure tones varying in intensity, or lights varying in luminance. A broad range of senses has been studied, including vision, hearing, touch (including skin and enteric perception), taste, smell, or the sense of time.

Psychoanalysis

Psychoanalysis is a family of psychological theories and methods based on the pioneering work of Sigmund Freud. As a result of talking with his patients, he came to believe that psychopathologies stemmed from culturally unacceptable, thus repressed and unconscious, desires and fantasies of a sexual nature. Moreover he stressed the presumed value of dreams as sources of insight into unconscious desires. The analyst's goal is thus aimed to discover connections among the unconscious components of patients' mental processes. This ought to help liberate the patient from unexamined or unconscious barriers of transference and resistance, that is, past patterns of relating that are no longer serviceable or that inhibit freedom.

Carl G. Jung's unique and broadly influential approach to psychology emphasized understanding the psyche through exploring the worlds of dreams, art, mythology, world religion and philosophy. Jung emphasized the importance of balance and harmony. He cautioned that modern humans rely too heavily on science and logic and would benefit from integrating spirituality and appreciation of the unconscious realm. His Collective unconscious refers to a human experience reservoir, a kind of knowledge, we are all born with but can never be directly conscious of. His Archetypes represent generic, idealized models of personalities with unlearned tendencies to experience things in a certain way.

COMPARATIVE PSYCHOLOGY AND BEHAVIORISM

This school of thought represents a movement in psychology that advocates the use of strict experimental procedures to study observable behaviors (or responses) in relation to environment (or stimuli). Behaviorists regard learning and environmental conditions as dominant over the effects of genetics and heredity. Concepts, such as learning or memory are explored using artificial tasks in strictly controlled environments (e.g., Skinner Box, Morris' water maze, Radial arm maze).

Comparative psychology viewed science less as a framework for explaining events but rather for predicting and controlling them. Adoption of the observational techniques that had proven useful in the study of the behavior of animals was viewed as the best way to advance the scientific study of human beings. The external environment represented the main determining influence on behavior and its manipulation held the most promise for improving the condition of human beings. The concept of a tabula rasa was revived.

A common tenet held that the complex phenomena of behavior could be explained by and reduced to clear, distinct, reflexive units of behavior.

William James described psychology as "the science of mental life." As a rigid application of the scientific method is only possible for learned behaviors, he felt that our studies should foremost focus on studying those.

Ivan P. Pavlov realized that behavior changes when an organism comes to associate one stimulus with another. Specifically, a reflexive or automatic response transfers from one stimulus to another - Classical Conditioning.

Edward L. Thorndike Using a Puzzle Box, he demonstrated that behaviors that are followed by a positive outcome are often repeated, while those followed by a negative outcome or none at all are extinguished (i.e., Operant Conditioning). He postulated the Law of effect where behavior changes when the results of an action proved of consequence (e.g. rewarding, noxious).

John B. Watson stimulus-response theory of psychology, emotional reactions are learned in much the same way as other skills

B.F. Skinner contends that nearly all behavior is shaped by complex patterns of reinforcement in a person's environment. Skinner Box: An animal placed inside the box is rewarded with a small bit of food each time it makes the desired response, such as pressing a lever or pecking a key. A device outside the box records the animal's responses.

Ethology

"When, as by a miracle, the lovely butterfly bursts from the chrysalis full-winged and perfect,... it has, for the most part, nothing to learn, because its little life flows from its organization like melody from a music box" -- Douglas A. Spalding, 1873

Ethology, the biological study of behavior under naturalistic conditions, emerged out of diverse roots by individuals largely trained in a zoological tradition. It aimed to provide a more comprehensive study of animal behavior than previously available. Where comparative psychology studies the behavior of animals in the context of what is known about human psychology, ethology explores animal behavior in the context of animal anatomy and physiology. Ethologists specifically considered the significance of inherited elements of behavior alongside learned components. Influenced by evolutionary biologists such as Thomas Hunt Morgan and others, many aspects of behavior may indeed be accessible to selection. Charles Otis Whitman examined the importance of phylogeny to the origins of behavior. Oskar Heinroth rather focused on how the adaptive value of behavior may explain behavioral differences among taxa. Experiments, in which he raised young birds separate from other individuals of their own species, demonstrated that many species-specific behaviors were nonetheless present. Interest in genetically-programmed elements (instincts) were contrasted to those subject to learning (e.g., imprinting). Using a comparative approach ethologists are thus interested to search for the biological basis of behavior (including human behavior) by comparing specific activities to those of close relatives.

Karl von Frisch studied the senses of honey bees, identified their mechanisms of communication, and showed their sensitivity to ultraviolet and polarized light. He unlocked the basic secrets of the forager's waggle dance, research which continues to yield new insights to this day. Techniques developed by behaviorists, such as classical and operant conditioning, have proven valuable to ethologists. Animals trained to respond to a particular stimulus in the appropriate behavioral context can demonstrate their ability to discriminate between different types of stimulus. For example, von Frisch used classical conditioning to show that fish could hear, and bees were able to perceive colors.

Konrad Lorenz' work included studies of imprinting (i.e., the capacity to learn specific types of information at certain critical periods in development) and instinct (i.e., the capacity of an animal to complete a complex behavior in response to a first-time encounter with a given situation). Filial imprinting refers to the case where young of a species need to acquire the image of what their parents look like (e.g. ducks). Instinctive behavior can usually be elicited by a limited combination of complex environmental cues. Such a sign stimulus refers to a particular subset of features or qualities of a natural stimulus or from complex environmental cues that is sufficient to elicit a specific behavioral response. They noted that different individuals produce identical behavioral response to specific sign stimuli - Motor patterns (or Fixed Action Pattern, FAP). Once initiated, FAPs continue until completion. A releasing mechanism is a functionally organized, neural circuit that recognizes a specific releaser and produces the appropriate response. Due to their stereotype, behaviors can be cataloged in form of an Ethogram (i.e., a comprehensive compilation of the behavioral repertory of a species, a catalog of actions as complete and precise as possible). An internal factor that determines how likely an animal is to perform a behavior (i.e., Action-Specific Energy).

The causation of behavior is viewed in the context of a metaphor - the Psychohydraulic model. Action specific energy is akin to a liquid building up in a vessel. A high level of action specific energy results when a behavior has not been performed in some time. Pressure from the water pushes onto a piston which is also pulled by sign stimuli in the environment, represented by weights. As pressure mounts the piston finally opens a valve that allows water to flow out (representing the performance of the behavior) Example: An animal that has had a good meal will have a low drive to eat more (i.e., little water in the apparatus). As it goes for a while without eating, the drive (i.e., hunger) builds up. If food is then encountered, the animal will try to consume it. If no good food is encountered, less preferred items will be accepted. This model has faced much criticism as performance of many behaviors appear to increase the tendency to perform it again (i.e. aggression), rather than bring about a decrease

Three Ethologists shared the 1973 Nobel Price for Physiology or Medicine in recognition of their work over the previous 30 years as founders of the field of ethology.

Karl von Frisch social behavior of bees

Konrad Z. Lorenz studied the behavior of geese, ravens, and ducks:

Niko Tinbergen behavioral studies of gulls and stickle backs. "Man likes to consider himself a rationale being; he not only fails to recognize his instincts, he even scoffs at instincts and considers them inferior. This attitude makes him fall victim to similarity between man and animals; he recognizes himself only too often in an animal. Also, he has rather a higher opinion of instincts; for instance, he recognizes that they are at the bottom of his subjective experiences. And he has doubts about the all-controlling power of reason, for he sees its limits too often."

Behavioral Genetics

Behavioral Genetics explores the hereditability of behavioral traits. More recently, this has given rise to a largely quantitative emphasis on using molecular genetic techniques in order to isolate individual genes that play a role in behavior, neural and behavioral pathologies, or individual human characteristics.

Each inherited feature of an organism is determined by a section of genetic code - a gene which represents the smallest hereditary unit. The genome refers to one individual's entire set of hereditary information and includes genes as well as non-coding sequences of DNA. The genomes for any two individuals, except for those of identical twins or cloned animals, are unique. Such differences (i.e., polymorphism) produce different versions of the same genes, called alleles. Despite these differences in genetic makeup, we also find major commonalities across much of the code shared by all forms of life. The specific genetic makeup of an individual is called its genotype. A trait refers to a specific characteristic or feature of an organism (although in psychology the term may be used to represent a component of personality). The term phenotype represents the specific manifestation of a trait or its physical appearance and constitution and refers to the actual state of that trait (e.g., the trait eye color has the phenotypes blue, brown and hazel). The individual's phenotype emerges as interaction of genetic information with environmental conditions.

Sir Francis Galton (February 16, 1822 – January 17, 1911), a younger relative of Charles Darwin, was an English explorer, geographer, and statistician. Intrigued by Darwin's concept of natural selection and with a keen interest in the breeding of domesticated animals, Galton explored variation in human populations and its implications. He became interested at first in the question of whether human ability was hereditary. He presented evidence for the inheritance of a range of abilities - "a man's natural abilities are derived by inheritance, under exactly the same limitations as are the form and physical features of the whole organic world". He also proposed adoption studies, including trans-racial adoption studies, to separate out the effects of heredity and environment. He was instrumental in the development of quantitative methods to test human intelligence, statistical correlation, fingerprint analysis. self-report questionnaires, and word-association studies. In a darker side, he became a vociferous proponent of eugenics where he proposed to improve the human race via selective breeding of those individuals who managed to score high marks on intelligence tests.

BEHAVIORAL ECOLOGY

Behavioral ecology is the study of the ecological and evolutionary basis for animal behavior. Primarily concerned with ultimate (evolutionary) consequences, researchers aim to identify phylogenetic contingencies within behavior (i.e., to what degree behavior is inherited from its ancestors) and to examine its functional significance (i.e., how the behavior impacts the bearers reproductive success). Behavioral ecologists contend that in order to minimize ad-hoc explanations arising from anecdotal observations a rigorous examination of the adaptive consequences of a behavior are essential. Specifically, it explores what a trait is good for in an evolutionary context.

Optimization theory is concerned with assessing the success of behavior. It attempts to identify those behavioral strategies that offer the highest return under a given set of conditions using a cost/benefit analysis. In this context success or fitness is judged by considering the number of offspring that the bearer of the behavior would contribute to the next generation relative to its suboptimal conspecifics. Optimization theory, such as optimal foraging theory, states that individuals ought to adopt strategies that allow it to consume the most energy while expending the least amount of energy. Evolutionary game theory asserts that the success of strategies should not only consider its inherent benefits, as the actual success depends on what other strategies exist in the population. An evolutionarily stable strategy is a strategy which when adopted by all members cannot be beaten out by any invading, mutant strategy.

Sociobiology

Sociobiology aims to explain behavior in all species by considering the evolutionary advantages of social behaviours. Sociobiology is based on the idea that all animals, including humans, will tend to act in ways to improve their own reproductive success, and that this will result in social processes conducive to genetic fitness. For example, cultural practices and values related to sexuality are viewed as shaped by natural selection (e.g., societal rules for incest avoidance). The concept has been criticized for its reliance on innate predispositions over learning, and by the view that presents situations are not representative of historical conditions.

The selfish genes hypothesis states that the individual is not as important to the population as its genes. Kin selection refers to changes in gene frequency across generations when biological relatives influence the reproductive success of each other. Thus, a gene that prompts behavior which enhances the fitness of relatives but lowers that of the individual displaying the behavior, may nonetheless increase in frequency, as relatives often share the same genes. The enhanced fitness of relatives can at times more than compensate for the fitness loss incurred by the individuals displaying the behavior.

The Scientific Method

The Scientific method refers to a body of techniques for investigating phenomena, acquiring new knowledge, or correcting and integrating previous knowledge. To be termed scientific, a method of inquiry must be based on gathering observable, empirical and measurable evidence subject to specific principles of reasoning. A scientific method consists of the collection of data through observation and experimentation, and the formulation and testing of hypotheses

History

Socrates developed logic as a way of thinking and speaking which would let you prove that a certain statement was or was not true. Plato (Socrates' student) continued this idea, and Plato's student Aristotle began to apply logic to the natural world in which arguments were used instead of experiments. Aristotle, dissatisfied with the knowledge that was imparted when the researcher moves from a set of specific facts to a general conclusion, rather relied on deductive reasoning. For centuries the library of Alexandria served as a center for scientific exploration. Following Roman conversion to Christianity its influence in the Western world declined around 600AD but its contents was preserved with the rise of islam where explorations of the natural world prospered. The experimental scientific method was developed by Alhazen, who in 1021AD used experimentation and mathematics to obtain the results in his Book of Optics. In the medieval Western world (600-1000) faith and the supernatural reigned supreme over the value of human reason and science. Giving rise to the Renaissance, the grip of church doctrine weakened after the 12th century with a variety of influences, including the black death, the influx of scholarly literature from the islamic library of Cordoba, and the printing press. Galileo Galilei (1564-1642) starts the Scientific Revolution and claims that church doctrine ought not to make claims about the natural world which can easily be shown to be false - and he pays the price. Renee Descartes (1596-1650) urges us to never accept anything as true that is not known clearly to be such. William Harvey (1578-1657) phrases the Scientific Method.

A method for knowing

The scientific method is the process by which scientists construct an accurate (i.e., reliable, consistent and non-arbitrary) representation of the world. Scientists attempt to minimize the influence of their biases on the outcome of an experiment. The most fundamental error is to mistake the hypothesis for an explanation of a phenomenon, without performing experimental tests. Sometimes "common sense" and "logic" tempt us into believing that no test is needed. There are numerous examples of this, dating from the Greek philosophers to the present day.The prime directive is to never accept anything as true that is not known clearly to be such. Towards this goal scientists progress through a series of structured steps.

Initial Observation tickles your interest in a particular topic

Phrase a Question that can be addressed with an experiment

Phrase Null Hypothesis as an explicit claim about the value of a specific parameter. What would you predict if the decision to turn left/right is random? In our example, the probability to turn right P(R) must equal the probability of turning left P(L). Because subjects have two mutually exclusive choices at the decision point we know that the probability of turning either left or right P(R or L) = P(R) + P(L) = 1, or P(R) = P(L) = 0.5. Thus, we can make an explicit claim about the probability that a subject turns right or left at the decision point given the choice is random. Our null hypothesis is that P(R) = 0.5 (or, equivalently, that P(L) = 0.5).

Directional vs. non-directional hypotheses

Rejection regions and one-tailed vs. two-tailed tests

Phrase Alternate Hypothesis which simply states that our null hypothesis is not true, or P(R) 0.5 (or, equivalently, P(L) 0.5).

Consider issues relating to the Treatment Effect

Decide what size of a treatment effect would be considered "significant".

Decide on sample size needed to detect a treatment effect of this size (i.e., power).

Decide on framework (experiment) for distinguishing between the two hypotheses. This allows us to attach a sense of likelihood to the outcomes associated with a particular outcome under the null hypothesis. Scientists must minimize their own biases with respect to a specific outcome and their preferences may not bias the results or their interpretation.

Consider the type of data that the experiment will produce and choose a statistical test that allows you to assess to what degree the experiment's outcome may be the result of chance alone. This step involves the computation of a specific test statistic. You are bound to retain the null hypothesis unless you have good evidence that it is not compatible with the outcome of the experiment. A comparison of the test statistic to a critical value allows us to either retain our null hypothesis or reject it in favor of the alternative hypothesis.

Choose acceptable level for falsely rejecting the null hypothesis (i.e., p-value).

Determine critical values for the sample statistic. We fail to reject the null hypothesis if our test statistic does not exceed the critical value and we reject the null hypothesis in favor of the alternative if the test statistic exceeds the critical value.

Conduct Experiment in randomized, double-blind fashion with the inclusion of appropriate controls.

Un-blind samples. Match each particular outcome with their respective treatments.

During Data Analysis all data must be handled in the same way. There is always the temptation to more closely inspect the validity of data points which do not support the scientist's expectations, while data which do agree with those expectations may not be checked as carefully.

Draw Conclusions

Communicate Results

Terminology

The general public often critically misinterprets terms used by scientists such as hypothesis, theory and scientific law. A hypothesis is any educated hunch phrased to explain an observed phenomenon so it becomes testable by an experiment. Even the most hair-brained ideas can be phrased into a hypothesis. Model or principle is reserved for situations that have at least limited validity. For example, Bohr's model of the atom is formed as an analogy to the solar system, where electrons move in circular orbits around the nucleus. This is not an accurate depiction of what an atom "looks like," but the model is to some degree able to represent the energies of electrons in a hydrogen atom. The term Scientific Law is reserved for a concise explanation of a simple set of actions that is accepted to be true and universal.

Scientific laws are similar to mathematical postulates. The latter really no longer need complex external proofs or experimental verification based upon the fact that they have always been observed to be true. Examples include the law of gravity, or the laws of thermodynamics. Both, laws and theories can be used to make predictions about the outcome of future events. Like a law, a theory is not any hair-brained idea that popped into somebody's head - it is rather an understanding of something that has been well documented, supported by overwhelming scientific evidence, has stood up to exhaustive rigorous testing, and is generally accepted as being true beyond reasonable doubt. Whereas laws govern a single action, a series of related phenomena can only be represented by a theory that describes the fundamental properties and relationships in a complex system.

As with an automobile used as a transportationd device, improvements are sometimes made to one or more component parts (e.g., a new set of spark plugs), but the function of the automobile as a whole remains unchanged. Similarly, individual components of a theory can be changed or improved upon, without changing the overall truth of the theory as a whole. Examples of scientific theories include the theory of evolution, the theory of relativity, and quantum theory. Although scientists continue to attempt to make its components more elegant, concise, or all-encompassing, they are seldom, if ever, entirely replaced.

Ockham's Razor: "Pluralitas non est ponenda sine neccesitate, which translates as "entities should not be multiplied unnecessarily -- William of Ockham, 14th century. It reminds us to keep things simple. In cases where we have competing theories describing the same phenomenon we are bound to go with the most parsimoneous explanation first. It does not mean that the simplest theory will be correct, it just focuses on priorities.

Tautological Arguments are a needless repetition of an idea in different words or a logical statement that is necessarily true because it includes all possibilities. It is logically true irrespective of whether the underlying statement is factually true or false because no alternate hypothesis is possible.

NATURAL ENVIRONMENT

The natural environment encompasses all living and non-living things occurringnaturally on Earth or some region there of. It is an environment that encompasses the interaction of all living species. Climate, weather, and natural resources that affect human survival and economic activity. The concept of the *natural environment* can be distinguished by components:

- Complete ecological units that function as natural systems without massive civilized human intervention, including all vegetation, microorganisms, soil,rocks, atmosphere, and natural phenomena that occur within their boundaries

- Universal natural resources and physical phenomena that lack clear-cut boundaries, such as air, water, and climate, as well as energy, radiation, electric charge, and magnetism, not originating from civilized human activity

In contrast to the natural environment is the built environment. In such areas where man has fundamentally transformed landscapes such as urban settings and agricultural land conversion, the natural environment is greatly modified and diminished, with a much more simplified human environment largely replacing it. Even events which seem less extreme such as hydroelectric dam construction, or solar array construction in the desert, the natural environment is substantially altered.

It is difficult to find *absolutely natural* environments, and it is common that the naturalness varies in a continuum, from ideally 100% natural in one extreme to 0% natural in the other. More precisely, we can consider the different aspects or components of an environment, and see that their degree of naturalness is not uniform. If, for instance, we take an agricultural field, and consider the mineralogic composition and the structure of its soil, we will find that whereas the first is quite similar to that of an undisturbed forest soil, the structure is quite different.

Natural environment is often used as a synonym for habitat. For instance, when we say that the natural environment of giraffes is the savanna.

COMPOSITION

Earth science generally recognizes 4 spheres, the lithosphere, the hydrosphere, the atmosphere, and the biosphere as correspondent to rocks, water, air, and life. Some scientists include, as part of the spheres of the Earth, the cryosphere(corresponding to ice) as a distinct portion of the hydrosphere, as well as the pedosphere(corresponding to soil) as an active and intermixed sphere. Earth science (also known as geoscience, the geosciences or the Earth Sciences), is an all-embracing term for the sciences related to the planet Earth. There are four major disciplinesin earth sciences, namely geography, geology, geophysics and geodesy. These major disciplines use physics, chemistry, biology, chronology and mathematics to build a qualitative and quantitative understanding of the principal areas or *spheres*of the Earth.

GEOLOGICAL ACTIVITY

The Earth's crust, or lithosphere, is the outermost solid surface of the planet and is chemically and mechanically different from underlying mantle. It has been generated greatly by igneous processes in which magma cools and solidifies to form solid rock. Beneath the lithosphere lies the mantle which is heated by the decay of radioactive elements. The mantle though solid is in a state of rheic convection. This convection process causes the lithospheric plates to move, albeit slowly. The resulting process is known as plate tectonics. Volcanoes result primarily from the melting of subducted crust material or of rising mantle at mid-ocean ridges and mantle plumes.

WATER ON EARTH

Oceans

An ocean is a major body of saline water, and a component of the hydrosphere. Approximately 71% of the Earth's surface (an area of some 362 million square kilometers) is covered by ocean, a continuous body of water that is customarily divided into several principal oceans and smaller seas. More than half of this area is over 3,000 meters (9,800 ft) deep. Average oceanic salinity is around 35 parts per thousand (ppt) (3.5%), and nearly all seawater has a salinity in the range of 30 to 38 ppt. Though generally recognized as several 'separate' oceans, these waters comprise one global, interconnected body of salt water often referred to as the World Ocean or global ocean. This concept of a global ocean as a continuous body of water with relatively free interchange among its parts is of fundamental importance tooceanography. The major oceanic divisions are defined in part by thecontinents, various archipelagos, and other criteria: these divisions are (in descending order of size) the Pacific Ocean, theAtlantic Ocean, the Indian Ocean, the Southern Ocean and the Arctic Ocean.

Rivers

A river is a natural watercourse, usuallyfreshwater, flowing toward an ocean, a lake, a sea or another river. In a few cases, a river simply flows into the ground or dries up completely before reaching another body of water. Small rivers may also be termed by several other names, including stream, creek and brook. In the United States a river is generally classified as a watercourse more than 60 feet (18 metres) wide. The water in a river is usually in a channel, made up of a stream bed between banks. In larger rivers there is also a wider floodplain shaped by waters over-topping the channel. Flood plains may be very wide in relation to the size of the river channel. Rivers are a part of the hydrological cycle. Water within a river is generally collected from precipitation through surface runoff, groundwater recharge, springs, and the release of water stored in glaciers and snowpacks.

Streams

A stream is a flowing body of water with a current, confined within a bed and stream banks. Streams play an importantcorridor role in connecting fragmented habitats and thus in conserving biodiversity. The study of streams and waterways in general is known as *surface hydrology.* Types of streams include creeks, tributaries, which do not reach an ocean and connect with another stream or river, brooks, which are typically small streams and sometimes sourced from a spring orseep and tidal inlets.

Lakes

A lake (from Latin *lacus*) is a terrain feature, a body of water that is localized to the bottom of basin. A body of water is considered a lake when it is inland, is not part of an ocean, is larger and deeper than a pond, and is fed by a river.

Natural lakes on Earth are generally found in mountainous areas, rift zones, and areas with ongoing or recent glaciation. Other lakes are found in endorheic basins or along the courses of mature rivers. In some parts of the world, there are many lakes because of chaotic drainage patterns left over from the last Ice Age. All lakes are temporary over geologic time scales, as they will slowly fill in with sediments or spill out of the basin containing them.

Ponds

A pond is a body of standing water, either natural or man-made, that is usually smaller than a lake. A wide variety of man-made bodies of water are classified as ponds, including water gardens designed for aesthetic ornamentation, fish pondsdesigned for commercial fish breeding, and solar ponds designed to store thermal energy. Ponds and lakes are distinguished from streams via current speed. While currents in streams are easily observed, ponds and lakes possess thermally driven micro-currents and moderate wind driven currents. These features distinguish a pond from many other aquatic terrain features, such as stream pools and tide pools.

Atmosphere, climate and weather

The atmosphere of the Earth serves as a key factor in sustaining the planetary ecosystem. The thin layer of gases that envelops the Earth is held in place by the planet's gravity. Dry air consists of 78% nitrogen, 21% oxygen, 1% argon and otherinert gases, such as carbon dioxide. The remaining gases are often referred to as trace gases, among which are the greenhouse gases such as water vapor, carbon dioxide, methane, nitrous oxide, and ozone. Filtered air includes trace amounts of many other chemical compounds. Air also contains a variable amount ofwater vapor and suspensions of water droplets and ice crystals seen as clouds. Many natural substances may be present in tiny amounts in an unfiltered air sample, including dust, pollen and spores, sea spray, volcanic ash, and meteoroids. Various industrial pollutants also may be present, such as chlorine (elementary or in compounds), fluorine compounds, elemental mercury, and sulphur compounds such as sulphur dioxide [SO_2].

The ozone layer of the Earth's atmosphere plays an important role in depleting the amount of ultraviolet (UV) radiation that reaches the surface. As DNA is readily damaged by UV light, this serves to protect life at the surface. The atmosphere also retains heat during the night, thereby reducing the daily temperature extremes.

Atmospheric layers

Principal layers

Earth's atmosphere can be divided into five main layers. These layers are mainly determined by whether temperature increases or decreases with altitude. From highest to lowest, these layers are:

- Exosphere: The outermost layer of Earth's atmosphere extends from the exobase upward, mainly composed of hydrogen and helium.
- Thermosphere: The top of the thermosphere is the bottom of the exosphere, called the exobase. Its height varies with solar activity and ranges from about 350–800 km (220–500 mi; 1,150,000–2,620,000 ft). The International Space Station orbits in this layer, between 320 and 380 km (200 and 240 mi).
- Mesosphere: The mesosphere extends from the stratopause to 80–85 km (50–53 mi; 262,000–279,000 ft). It is the layer where most meteors burn up upon entering the atmosphere.
- Stratosphere: The stratosphere extends from the tropopause to about 51 km (32 mi; 167,000 ft). The stratopause, which is the boundary between the stratosphere and mesosphere, typically is at 50 to 55 km (31 to 34 mi; 164,000 to 180,000 ft).
- Troposphere: The troposphere begins at the surface and extends to between 7 km (23,000 ft) at the poles and 17 km (56,000 ft) at the equator, with some variation due to weather. The troposphere is mostly heated by and nations in adjusting to the effects of global warming. Some examples of recent collaboration to address climate change and global warming include:
- The United Nations Framework Convention Treaty and convention on Climate Change, to stabilize greenhouse gas concentrations in the atmosphere at a level that would prevent dangerous anthropogenic interference with the climate system.
- The Kyoto Protocol, which is the protocol to the international Framework Convention on Climate Change treaty, again with the objective of reducing greenhouse gases in an effort to prevent anthropogenic climate change.

- The Western Climate Initiative, to identify, evaluate, and implement collective and cooperative ways to reduce greenhouse gases in the region, focusing on a market-based cap-and-trade system.

A significantly profound challenge is to identify the natural environmental dynamics in contrast to environmental changes not within natural variances. A common solution is to adapt a static view neglecting natural variances to exist. Methodologically, this view could be defended when looking at processes which change slowly and short time series, while the problem arrives when fast processes turns essential in the object of the study.

Climate

Climate encompasses the statistics of temperature, humidity, atmospheric pressure,wind, rainfall, atmospheric particle count and numerous other meteorologicalelements in a given region over long periods of time. Climate can be contrasted to weather, which is the present condition of these same elements over periods up to two weeks.

Climates can be classified according to the average and typical ranges of different variables, most commonly temperature and precipitation. The most commonly used classification scheme is the one originally developed by Wladimir Köppen. The Thornthwaite system, in use since 1948, incorporates evapotranspiration in addition to temperature and precipitation information and is used in studying animal species diversity and potential impacts of climate changes.

Weather

Weather is a set of all the phenomena occurring in a given atmospheric area at a given time. Most weather phenomena occur in the troposphere, just below the stratosphere. Weather refers, generally, to day-to-day temperature and precipitation activity, whereas climate is the term for the average atmospheric conditions over longer periods of time. When used without qualification, "weather" is understood to be the weather of Earth.

Weather occurs due to density (temperature and moisture) differences between one place and another. These differences can occur due to the sun angle at any particular spot, which varies by latitude from the tropics. The strong temperature contrast between polar and tropical air gives rise to the jet stream. Weather systems in the mid-latitudes, such as extratropical cyclones, are caused by instabilities of the jet stream flow. Because the Earth's axis is tilted relative to its orbital plane, sunlightis incident at different angles at different times of the year. On the Earth's surface, temperatures usually range ±40 C (100 F to "40 F) annually. Over thousands of years, changes in the Earth's orbit have affected the amount and distribution of solar energy received by the Earth and influence long-term climate

Surface temperature differences in turn cause pressure differences. Higher altitudes are cooler than lower altitudes due to differences in compressional heating. Weather forecasting is the application of science and technology to predict the state of the atmosphere for a future time and a given location. The atmosphere is a chaotic system, and small changes to one part of the system can grow to have large effects on the system as a whole. Human attempts to control the weather have occurred throughout human history, and there is evidence that civilized human activity such as agriculture and industry has inadvertently modified weather patterns.

Life

Evidence suggests that life on Earth has existed for about 3.7 billion years. All known life forms share fundamental molecular mechanisms, and based on these observations, theories on the origin of life attempt to find a mechanism explaining the formation of a primordial single cell organism from which all life originates. There are many different hypotheses regarding the path that might have been taken from simple organic molecules via pre-cellular life to protocells and metabolism.

Although there is no universal agreement on the definition of life, scientists generally accept that the biological manifestation of life is characterized by organization, metabolism, growth, adaptation, response to stimuli and reproduction. Life may also be said to be simply the characteristic state of organisms. In biology, the science of living organisms, "life" is the condition which distinguishes activeorganisms from inorganic matter, including the capacity for growth, functional activity and the continual change preceding death.

A diverse variety of living organisms (life forms) can be found in the biosphere on Earth, and properties common to these organisms—plants, animals, fungi, protists, archaea, and bacteria—are a carbon- and water-based cellular form with complex organization and heritable genetic information. Living organisms undergo metabolism, maintain homeostasis, possess a capacity to grow, respond to stimuli, reproduce and, through natural selection, adapt to their environment in successive generations. More complex living organisms can communicate through various means

Ecosystems

An ecosystem (also called as environment) is a natural unit consisting of all plants, animals and micro-organisms (biotic factors) in an area functioning together with all of the non-living physical (abiotic) factors of the environment.

Central to the ecosystem concept is the idea that living organisms are continually engaged in a highly interrelated set of relationships with every other element constituting the environment in which they exist. Eugene Odum, one of the founders of the science of ecology, stated: "Any unit that includes all of the organisms (ie: the "community") in a given area interacting with the physical environment so that a flow of energy leads to clearly defined trophic structure, biotic diversity, and material cycles (i.e.: exchange of materials between living and nonliving parts) within the system is an ecosystem."

The human ecosystem concept is then grounded in the deconstruction of the human/nature dichotomy, and the emergent premise that all species are ecologically integrated with each other, as well as with the abiotic constituents of their biotope.

A greater number or variety of species or biological diversity of an ecosystem may contribute to greater resilience of an ecosystem, because there are more species present at a location to respond to change and thus "absorb" or reduce its effects. This reduces the effect before the ecosystem's structure is fundamentally changed to a different state. This is not universally the case and there is no proven relationship between the species diversity of an ecosystem and its ability to provide goods and services on a sustainable level.

The term ecosystem can also pertain to human-made environments, such ashuman ecosystems and human-influenced ecosystems, and can describe any situation where there is relationship between living organisms and their environment. Fewer areas on the surface of the earth today exist free from human contact, although some genuine wilderness areas continue to exist without any forms of human intervention.

BIOMES

Biomes are terminologically similar to the concept of ecosystems, and areclimatically and geographically defined areas of ecologically similar climatic conditions on the Earth, such as communities of plants, animals, and soil organisms, often referred to *as* ecosystems. Biomes are defined on the basis of factors such as plant structures (such as trees, shrubs, and grasses), leaf types (such as broadleaf and needleleaf), plant spacing (forest, woodland, savanna), and climate. Unlike ecozones, biomes are not defined by genetic, taxonomic, or historical similarities. Biomes are often identified with particular patterns of ecological succession and climax vegetation.

Biogeochemical cycles

Global biogeochemical cycles are critical to life, most notably those of water,oxygen, carbon, nitrogen and phosphorus.

- The nitrogen cycle is the transformation of nitrogen and nitrogen-containing compounds in nature. It is a cycle which includes gaseous components.
- The water cycle, is the continuous movement of water on, above, and below the surface of the Earth. Water can change states among liquid, vapor, and ice at various places in the water cycle. Although the balance of water on Earth remains fairly constant over time, individual water molecules can come and go.
- The carbon cycle is the biogeochemical cycle by which carbon is exchanged among the biosphere, pedosphere, geosphere, hydrosphere, and atmosphere of the Earth.
- The oxygen cycle is the movement of oxygen within and between its three main reservoirs: the atmosphere, the biosphere, and the lithosphere. The main driving factor of the oxygen cycle is photosynthesis, which is responsible for the modern Earth's atmospheric composition and life.
- The phosphorus cycle is the movement of phosphorus through the lithosphere, hydrosphere, and biosphere. The atmosphere does not play a significant role in the movements of phosphorus, because phosphorus and phosphorus compounds are usually solids at the typical ranges of temperature and pressure found on Earth.

Wilderness

Wilderness is generally defined as a natural environment on Earth that has not been significantly modified by human activity. The WILD Foundation goes into more detail, defining wilderness as: "The most intact, undisturbed wild natural areas left on our planet - those last truly wild places that humans do not control and have not developed with roads, pipelines or other industrial infrastructure." Wilderness areas and protected parks are considered important for the survival of certainspecies, ecological studies, conservation, solitude, and recreation. Wilderness is deeply valued for cultural, spiritual, moral, and aesthetic reasons. Some nature writers believe wilderness areas are vital for the human spirit and creativity.

The word, "wilderness", derives from the notion of wildness; in other words that which is not controllable by humans. The word's etymology is from the Old English *wildeornes*, which in turn derives from *wildeor* meaning *wild beast* (wild + deor = beast, deer). From this point of view, it is the wildness of a place that makes it a wilderness. The mere presence or activity of people does not disqualify an area from being "wilderness." Many ecosystems that are, or have been, inhabited or influenced by activities of people may still be considered "wild." This way of looking at wilderness includes areas within which natural processes operate without very noticeable human interference.

Wildlife includes all non-domesticated plants, animals and other organisms. Domesticating wild plant and animal species for human benefit has occurred many times all over the planet, and has a major impact on the environment, both positive and negative. Wildlife can be found in all ecosystems. Deserts, rain forests, plains, and other areas—including the most developed urban sites—all have distinct forms of wildlife. While the term in popular culture usually refers to animals that are untouched by civilized human factors, most scientists agree that wildlife around the world is (now) impacted by human activities.

Challenges

It is the common understanding of *natural environment* that underliesenvironmentalism — a broad political, social, and philosophical movement that advocates various actions and policies in the interest of protecting what nature remains in the natural environment, or restoring or expanding the role of nature in this environment. While true wilderness is increasingly rare, *wild* nature (e.g., unmanaged forests, uncultivated grasslands, wildlife, wildflowers) can be found in many locations previously inhabited by humans.

Goals commonly expressed by environmental scientists include:

- Reduction and cleanup of pollution, with future goals of zero pollution;
- Cleanly converting non-recyclable materials into energy through direct combustion or after conversion into secondary fuels;
- Reducing societal consumption of non-renewable fuels;
- Development of alternative, green, low-carbon or renewable energy sources;
- Conservation and sustainable use of scarce resources such as water, land, and air;
- Protection of representative or unique or pristine ecosystems;
- Preservation of threatened and endangered species extinction;
- The establishment of nature and biosphere reserves under various types of protection; and, most generally, the protection of biodiversity and ecosystems upon which all human and other life on earth depends.

Criticism

In some cultures the term environment is meaningless because there is no separation between people and what they view as the natural world, or their surroundings. Specifically in the U.S., many native cultures do not recognize the "environment", or see themselves as environmentalists.

2

History of Animal Behavior Study

The origins of the scientific study of animal behaviour lie in the works of various European thinkers of the 17th to 19th centuries, such as British naturalists John Ray and Charles Darwinand French naturalist Charles LeRoy. These individuals appreciated the complexity and apparent purposefulness of the actions of animals, and they knew that understanding behaviour demands long-term observations of animals in their natural settings. At first, the principal attraction of natural history studies was to confirm the ingenuity of God. The publication of Darwin's On the Origin of Species in 1859 changed this attitude. In his chapter on instinct, Darwin was concerned with whether behavioral traits, like anatomical ones, can evolve as a result of natural selection. Since then, biologists have recognized that the behaviours of animals, like their anatomical structures, are adaptations that exist because they have, over evolutionary time (that is, throughout the formation of new species and the evolution of their special characteristics), helped their bearers to survive and reproduce.

Furthermore, humans have long appreciated how beautifully and intricately the behaviours of animals are adapted to their surroundings. For example, young birds that possess camouflaged colour patterns for protection against predators will freeze when the parent spots a predator and calls the alarm. Darwin's achievement was to explain how such wondrously adapted creatures could arise from a process other than special creation. He showed that adaptation is an inexorable result of four basic characteristics of living organisms:

1. There is variation among individuals of the same species. Even closely related individuals, such as parent and offspring or sibling and sibling, differ considerably. Familiar human examples include differences in facial features, hair and eye colour, height, and weight.

2. Many of these variations areinheritable-that is, offspring resemble their parents in many traits as a result of the genes they share.
3. There are differences in numbers of surviving offspring among parents in every species. For example, one female snapping turtle (family Chelydridae) may lay 24eggs; however, only 5 may survive to adulthood. In contrast, another female may lay only 18 eggs, with 1 of her offspring surviving to adulthood.
4. The individuals that are best equipped to survive and reproduce perpetuate the highest frequency of genes to descendant populations. This is the principle known colloquially as "survival of the fittest," where fitness denotes an individual's overall ability to pass copies of his genes on to successive generations. For example, a woman who rears six healthy offspring has greater fitness than one who rears just two.
5. An inevitable consequence of variation, inheritance, and differential reproduction is that, over time, the frequency of traits that render individuals better able to survive and reproduce in their present environment increases. As a result, descendant generations in a population resemble most closely the members of ancestral populations that were able to reproduce most effectively. This is the process of natural selection.
6. The natural history approach of Darwin and his predecessors gradually evolved into the twin sciences of animal ecology, the study of the interactions between an animal and its environment, and ethology, the biological study of animal behaviour. The roots of ethology can be traced to the late 19th and early 20th centuries, when scientists from several countries began exploring the behaviours of selected vertebratespecies: dogs by the Russian physiologist Ivan Pavlov; rodents by American psychologists John B. Watson, Edward Tolman, and Karl Lashley; birds by American psychologist B.F. Skinner; and primates by German American psychologist Wolfgang Köhler and American psychologist Robert Yerkes. The studies were carried out in laboratories, in the case of dogs, rodents and pigeons, or in artificial colonies and laboratories, in the case of primates. These studies were oriented toward psychological and physiological questions rather than ecological or evolutionary ones.

It was not until the 1930s that field naturalists-such as English biologist Julian Huxley, Austrian zoologist Konrad Lorenz, and Dutch-born British zoologist and ethologist Nikolaas Tinbergenstudying birds and Austrian zoologist Karl von Frisch and American entomologist William Morton Wheeler examining insects-gained prominence and returned to broadly biological studies of animal behaviour. These individuals, the founders of ethology, had direct experience with the richness of the behavioral repertoires of animals living in their natural surroundings. Their "return to nature" approach was, to a large extent, a reaction against the tendency prevalent among psychologists to study just a few behavioral phenomena observed in a handful of species that were kept in impoverished laboratory environments.

The goal of the psychologists was to formulate behavioral hypotheses that claimed to have general applications (e.g., about learning as a single, all-purpose phenomenon). Later they would proceed using a deductive approach by testing their hypotheses through experimentation on captive animals. In contrast, the ethologists advocated an inductive approach, one that begins with observing and describing what animals do and then proceeds to address a general question: Why do these animals behave as they do? By this they meant "How do the specific behaviours of these animals lead to differential reproduction?" Since its birth in the 1930s, the ethological approach-which stresses the direct observation of a broad array of animal species in nature, embraces the vast variety of behaviours found in the animal kingdom, and commits to investigating behaviour from a broad biological perspective-has proved highly effective.

One of Tinbergen's most important contributions to the study of animal behaviour was to stress that ethology is like any other branch of biology, in that a comprehensive study of any behaviour must address four categories of questions, which today are called "levels of analysis," including causation, ontogeny, function, and evolutionary history. Although each of these four approaches requires a different kind of scientific investigation, all contribute to solving the enduring puzzle of how and why animals, including humans, behave as they do. A familiar example of animal behaviour-a dogwagging its tail-serves to illustrate the levels of analysis framework. When a dog senses the approach of a companion (dog or human), it stands still, fixates on the approaching individual, raises its tail, and begins swishing it from side to side. Why does this dog wag its tail? To answer this general question, four specific questions must be addressed.

HISTORY OF ZOOLOGY

The history of zoology before Charles Darwin's 1859 theory of evolution traces the organized study of the animal kingdom from ancient to modern times. Although the concept of zoology as a single coherent field arose much later, systematic study of zoology is seen in the works of Aristotle and Galen in the ancient Greco-Roman world. This work was developed in the Middle Ages by Islamic medicine and scholarship, and in turn their work was extended by European scholars such as Albertus Magnus.

During the European Renaissance and early modern period, zoological thought was revolutionized in Europe by a renewed interest in empiricism and the discovery of many novel organisms. Prominent in this movement were the anatomist Vesalius and the physiologist William Harvey, who used experimentation and careful observation, and naturalists such as Carl Linnaeus and Buffon who began to classify the diversity of life and the fossil record, as well as the development and behavior of organisms. Microscopy revealed the previously unknown world of microorganisms, laying the groundwork for cell theory. The growing importance of natural theology, partly a response to the rise of mechanical philosophy, encouraged the growth of natural history (although it entrenched the argument from design).

Over the 18th and 19th centuries, zoology became increasingly professional scientific disciplines. Explorer-naturalists such as Alexander von Humboldt investigated the interaction between organisms and their environment, and the ways this relationship depends on geography-laying the foundations for biogeography, ecology and ethology. Naturalists began to reject essentialism and consider the importance of extinction and the mutability of species. Cell theory provided a new perspective on the fundamental basis of life. These developments, as well as the results from embryology and paleontology, were synthesized in Charles Darwin's theory of evolution by natural selection. In 1859, Darwin placed the theory of organic evolution on a new footing, by his discovery of a process by which organic evolution can occur, and provided observational evidence that it had done so.

Pre-Scientific Zoology

The earliest humans must have had and passed on knowledge about animals to increase their chances of survival. This may have included unsystematic knowledge of human and animal anatomy and aspects of animal behavior (such as migration patterns). People learnt more about animals with the Neolithic Revolution about 10,000 years ago. Humans domesticated animals as people became pastoralists and then farmers instead of hunter-gatherers in civilisations such as those of ancient Egypt.

Ancient Eastern Cultures

The ancient cultures of Mesopotamia, the Indian subcontinent, and China, among others, produced renowned surgeons and students of the natural sciences such as Susruta and Zhang Zhongjing, reflecting independent sophisticated systems of natural philosophy. Taoist philosophers, such as Zhuangzi in the 4th century BC, expressed ideas related to evolution, such as denying the fixity of biological species and speculating that species had developed differing attributes in response to differing environments. The ancient Indian Ayurveda tradition independently developed the concept of three humours, resembling that of the four humours of ancient Greek medicine, though the Ayurvedic system included further complications, such as the body being composed of five elements and seven basic tissues. Ayurvedic writers also classified living things into four categories based on the method of birth (from the womb, eggs, heat & moisture, and seeds) and explained the conception of a fetus in detail. They also made considerable advances in the field of surgery, often without the use of human dissection or animal vivisection. One of the earliest Ayurvedic treatises was the Sushruta Samhita, attributed to Sushruta in the 6th century BC. It was also an early materia medica, describing 700 medicinal plants, 64 preparations from mineral sources, and 57 preparations based on animal sources. However, the roots of modern zoology are usually traced back to the secular tradition of ancient Greek philosophy.

Ancient Greek Traditions

The pre-Socratic philosophers asked many questions about life but produced little systematic knowledge of specifically zoological interest-though the attempts of the atomists to explain life in purely physical terms would recur periodically through the history of zoology. However, the medical theories of Hippocrates and his followers, especially humorism, had a lasting impact.

Aristotle

The philosopher Aristotle created the science of biology, basing its theory on both his metaphysicalprinciples and on observation. He proposed theories for the processes of metabolism, temperature regulation, information processing, embryonic development and inheritance. He made detailed observations of nature, especially the habits and attributes of animals in the sea at Lesbos. He classified 540 animal species, and dissected at least 50. Aristotle, and nearly all Western scholars after him until the 18th century, believed that creatures were arranged in a graded scale of perfection rising from plants on up to humans: the scala naturae or Great Chain of Being.

Hellenistic Zoology

A few scholars in the Hellenistic period under the Ptolemies-particularly Herophilus of Chalcedon and Erasistratus of Chios-amended Aristotle's physiological work, even performing experimental dissections and vivisections. Claudius Galen became the most important authority on medicine and anatomy. Though a few ancient atomists such as Lucretius challenged the teleological Aristotelian viewpoint that all aspects of life are the result of design or purpose, teleology (and after the rise of Christianity, natural theology) remained central to biological thought until the 18th and 19th centuries.

Medieval and Islamic Zoology

The decline of the Roman Empire led to the disappearance or destruction of much knowledge, though physicians still incorporated many aspects of the Aristotelian tradition into training and practice. In Byzantiumand the Islamic world, many of Aristotle's works were translated into Arabic and commented upon by scholars such as Avicenna and Averroes.

Medieval Muslim physicians, scientists and philosophers made significant contributions to zoological knowledge between the 8th and 13th centuries during the Islamic Golden Age. The Afro-Arab scholar al-Jahiz (781-869) described early evolutionary ideas such as the struggle for existence. He also introduced the idea of a food chain, and was an early adherent of environmental determinism.

During the High Middle Ages, a few European scholars such as Hildegard of Bingen, Albertus Magnus and Frederick II expanded the natural history canon. Magnus's De animalibus libri XXVI was one of the most extensive studies of zoological observation published before modern times.

RENAISSANCE AND EARLY MODERN

From Anatomy to Systematic Taxonomy

The Renaissance was the age of collectors and travellers, when many of the stories were actually demonstrated as true when the living or preserved specimens were brought to Europe. Verification by collecting of things, instead of the accumulation of anecdotes, then became more common, and scholars developed a new faculty of careful observation. The Renaissance brought expanded interest in both empirical natural history and physiology. In 1543, Andreas Vesalius inaugurated the modern era of Western medicine with his seminal human anatomy treatise De humani corporis fabrica, which was based on dissection of corpses. Vesalius was the first in a series of anatomists who gradually replaced scholasticism with empiricism in physiology and medicine, relying on first-hand experience rather than authority and abstract reasoning. Bestiaries-a genre that combines both the natural and figurative knowledge of animals-also became more sophisticated. Conrad Gessner great zoological work, Historiae animalium, appeared in four volumes, 1551-1558, at Zürich, a fifth being issued in 1587. His works were the starting-point of modern zoology. Other major works were produced by William Turner, Pierre Belon, Guillaume Rondelet, and Ulisse Aldrovandi. Artists such as Albrecht Dürerand Leonardo da Vinci, often working with naturalists, were also interested in the bodies of animals and humans, studying physiology in detail and contributing to the growth of anatomical knowledge.

In the 17th century, the enthusiasts of the new sciences, the investigators of nature by means of observation and experiment, banded themselves into academies or societies for mutual support and discourse. The first founded of surviving European academies, the Academia Naturae Curiosorum (1651) especially confined itself to the description and illustration of the structure of plants and animals; eleven years later (1662) the Royal Society of London was incorporated by royal charter, having existed without a name or fixed organisation for seventeen years previously (from 1645).

A little later the Academy of Sciences of Paris was established by Louis XIV, later still the Royal Society of Sciences in Uppsala was founded. Systematizing, naming and classifying dominated zoology throughout much of the 17th and 18th centuries. Carl Linnaeuspublished a basic taxonomy for the natural world in 1735 (variations of which have been in use ever since), and in the 1750s introduced scientific names for all his species. While Linnaeus conceived of species as unchanging parts of a designed hierarchy, the other great naturalist of the 18th century, Georges-Louis Leclerc, Comte de Buffon, treated species as artificial categories and living forms as malleable-even suggesting the possibility of common descent. Though he was writing in an era before evolution existed, Buffon is a key figure in the history of evolutionary thought; his "transformist" theory would influence the evolutionary theories of both Jean-Baptiste Lamarck and Charles Darwin.

Before the Age of Exploration, naturalists had little idea of the sheer scale of biological diversity. The discovery and description of new species and the collection of specimens became a passion of scientific gentlemen and a lucrative enterprise for entrepreneurs; many naturalists traveled the globe in search of scientific knowledge and adventure.

Extending the work of Vesalius into experiments on still living bodies (of both humans and animals), William Harvey investigated the roles of blood, veins and arteries. Harvey's De motu cordis in 1628 was the beginning of the end for Galenic theory, and alongside Santorio Santorio's studies of metabolism, it served as an influential model of quantitative approaches to physiology.

Impact of the Microscope

In the early 17th century, the micro-world of zoology was just beginning to open up. A few lensmakers and natural philosophers had been creating crude microscopes since the late 16th century, and Robert Hooke published the seminal Micrographia based on observations with his own compound microscope in 1665. But it was not until Antony van Leeuwenhoek's dramatic improvements in lensmaking beginning in the 1670s-ultimately producing up to 200-fold magnification with a single lens-that scholars discovered spermatozoa, bacteria, infusoria and the sheer strangeness and diversity of microscopic life. Similar investigations by Jan Swammerdam led to new interest in entomology and built the basic techniques of microscopic dissection and staining.

Debate over the flood described in the Bible catalyzed the development of paleontology; in 1669 Nicholas Steno published an essay on how the remains of living organisms could be trapped in layers of sediment and mineralized to produce fossils. Although Steno's ideas about fossilization were well known and much debated among natural philosophers, an organic origin for all fossils would not be accepted by all naturalists until the end of the 18th century due to philosophical and theological debate about issues such as the age of the earth and extinction.

Advances in microscopy also had a profound impact on biological thinking. In the early 19th century, a number of biologists pointed to the central importance of the cell. In 1838 and 1839, Schleiden and Schwann began promoting the ideas that (1) the basic unit of organisms is the cell and (2) that individual cells have all the characteristics of life, though they opposed the idea that (3) all cells come from the division of other cells. Thanks to the work of Robert Remak and Rudolf Virchow, however, by the 1860s most biologists accepted all three tenets of what came to be known as cell theory.

In Advance of on the Origin of Species

Up through the 19th century, the scope of zoology was largely divided between physiology, which investigated questions of form and function, and natural history, which was concerned with the diversity of life and interactions among different forms of life and between life and non-life. By 1900, much of these domains overlapped, while natural history (and its counterpart natural philosophy) had largely given way to more specialized scientific disciplines-cytology, bacteriology, morphology, embryology, geography, and geology. Widespread travel by naturalists in the early-to-mid-19th century resulted in a wealth of new information about the diversity and distribution of living organisms. Of particular importance was the work of Alexander von Humboldt, which analyzed the relationship between organisms and their environment (i.e., the domain of natural history) using the quantitative approaches of natural philosophy (i.e., physics and chemistry). Humboldt's work laid the foundations of biogeography and inspired several generations of scientists.

The emerging discipline of geology also brought natural history and natural philosophy closer together; Georges Cuvier and others made great strides in comparative anatomy and paleontology in the late 1790s and early 19th century. In a series of lectures and papers that made detailed comparisons between living mammals and fossil remains Cuvier was able to establish that the fossils were remains of species that had become extinct-rather than being remains of species still alive elsewhere in the world, as had been widely believed. Fossils discovered and described by Gideon Mantell, William Buckland, Mary Anning, and Richard Owen among others helped establish that there had been an 'age of reptiles' that had preceded even the prehistoric mammals. These discoveries captured the public imagination and focused attention on the history of life on earth.

The most significant evolutionary theory before Darwin's was that of Jean-Baptiste Lamarck; based on the inheritance of acquired characteristics (an inheritance mechanism that was widely accepted until the 20th century), it described a chain of development stretching from the lowliest microbe to humans.The British naturalist Charles Darwin, combining the biogeographical approach of Humboldt, the uniformitarian geology of Lyell, Thomas Malthus's writings on population growth, and his own morphological expertise, created a more successful evolutionary theory based on natural selection; similar evidence led Alfred Russel Wallace to independently reach the same conclusions. Charles Darwin's early interest in nature led him on a five-year voyage on HMS Beagle which established him as an eminent geologist whose observations and theories supported Charles Lyell's uniformitarian ideas, and publication of his journal of the voyage made him famous as a popular author. Puzzled by the geographical distribution of wildlife and fossils he collected on the voyage, Darwin investigated the transmutation of species and conceived his theory of natural selection in 1838. Although he discussed his ideas with several naturalists, he needed time for extensive research and his geological work had priority. He was writing up his theory in 1858 when Alfred Russel Wallace sent him an essay which described the same idea, prompting immediate joint publication of both of their theories. Darwin's On the Origin of Species, published on 24 November 1859, a seminal work of scientific literature, was to be the foundation of evolutionary biology.

The History of Behaviorism Psychology

Today the most commonly studied field of psychology is behaviorism. Behaviorism is the school of psychology where the goal is the prediction and control of behavior. Behaviorism today is a result of its philosophical roots and the work of major theorists. Three major theorists who contributed to behaviorism are John B. Watson, Edward Chace Tolman, and B.F. Skinner.

Prior to behaviorism, objective psychology was the main area of study in the field. Development in this field took place mainly in Russia by Ivan M. Sechenov. Objective psychology was psychology that insisted on studying only those things that are directly measurable. The most important concept that Sechenov emphasized was inhibition which he describes in his work Reflexes of the Brain. Inhibition is the reduction or cessation of activity caused by stimulation. This led him to conclude that all human behavior could be explained in terms of brain physiology, and can be reflexive. He also stated the only valid approach to the study of psychology involves the objective method of physiology.

Before the discovery of behaviorism, there were contributors that influenced behavioristic ideas. Vladimir M. Bechterev was a Russian psychologist who founded reflexology, and later claimed to be another founder of objective psychology. Reflexology is the study of human behavior that seeks to understand the relationship between environmental influences and overt behavior. His main goal was concentrated between environmental stimulus and behavior. Edward Thorndike, an American psychologist, also contributed to behaviorism's revolutionary ideas. His famous works on the learning theory lead to the development of operant conditioning within behaviorism. Thorndike developed the learning theory through studying animals which led to his ¿½law of effect¿½. His ¿½law of effect¿½ concluded that any behavior followed by pleasant consequences is likely to be continued, and any behavior followed by unpleasant consequences is likely to discontinue. Russian psychologist Ivan Pavlov also made major contributions to the field of behaviorism. He was most famous for his work in a type of learning called classical conditioning. He conducted his famous experiment with dogs and their digestive system. He developed terms from the experiment. Conditioned stimulus was a biologically neutral response, while a conditioned response was a neutral stimulus that elicits a fraction of the unconditioned response. Unconditioned reflex is innate and is triggered by the unconditioned stimulus. For example, in the dog study, the food powder in the dog's mouth will increase the saliva flow. Therefore the increase in salivation would be the unconditioned response. He concluded that conditioned reflexes, like natural reflexes could be explained in terms of physiology of the brain.

John B. Watson is founder of Behaviorism. Before this field of psychology was discovered, Watson discovered Russian Objective Psychology. Watson found that his ideas had a lot in common with those of the Russian psychologists. Both thought consciousness can cause behavior, rejected the idea of introspection and the explanation of behavior based on mentalism. However, unlike the Russian psychologists, Watson became less interested in the physiology of the brain and more on associating stimuli with responses. After studying these Russian psychologists, he viewed the goal of psychology as the prediction and control of behavior.

Watson developed many ideas during his research. He explained there are four types of learning. There is explicit learned behavior which consisted of speaking, writing, or playing baseball and explicit unlearned behavior, for example grasping, blinking, and sneezing. Implicit learned behavior for example, increased heart rate due to a particular fear, and implicit unlearned behavior such as glandular secretions or circulatory changes. Watson also felt that there are four methods of conducting research. The first method is observation which is naturalistic or experimentally controlled. The second method is the conditioned-reflex method, which came from Pavlov and Bechterev's works. The third method is testing, which emphasized taking behavior samples. The fourth method consists of verbal reports such as surveys or questionnaire's. He explained that all of these methods of studying research were effective.

Watson was very opinionated on particular issues. He emphasized that all humans inherit the emotions of fear, rage, and love. The main characteristics of emotions that adults develop are hate, pride, and jealousy. However, all of these characteristics are derived from fear, rage, and love. The mind-body relationship also remained important to Watson. He developed four different views on the mind-body relationship. The first view was the interactionist relationship, which simplified is the mind and body influencing each other. The second view was psychophysical parallelism. This meant that mental and bodily events are parallel with no interaction between the two. The third view was epiphenomenalism which explained that mental events are by-products of bodily events but do not cause behavior. Finally, the fourth was physical monism which completely rejected the existence of any mental events.

John B. Watson's work was a lasting influence in the field of psychology. He changed the whole goal of psychology. Psychology began strictly as an explanation of a state consciousness and shifted more toward predictions and controls of behavior. Therefore, he made overt behavior the main topic and concern in psychology. Two types of behaviorism have developed since Watson. Radical behaviorism is the belief that the explanation of behavior cannot be explained in terms of unobserved internal events, and if these events exist, then they should be ignored. The second type of behaviorism is methodological behaviorism. Methodological behaviorism accepts that overt behavior should be the main topic of psychology, but it should also consider internal causes of behavior including physiological and mental states.

Edward Chace Tolman was a major theorist who contributed to behaviorism. Tolman focused on another division of behaviorism which emphasized purposive behavior. Purposive behavior is directed toward a goal that terminates when the goal is attained. An example of purposive behavior is a rat running through a maze. He later translated purposive behavior into molar behavior, in order to contrast it with molecular behavior. Molecular behavior is a small section of behavior that is isolated for study. Tolman emphasized molar over molecular behavior.

During Tolman's experiments, he used intervening variables. Intervening variables occur between environmental and behavioral events. His use of intervening variables explained how the environment influenced behavior. An example of an intervening variable would be Tolman's cognitive map. A cognitive map is a mental representation of the environment. The cognitive map is a way of associating one landmark to another, which is useful when reading directions.

One of Tolman's most renowned experiments involved latent learning. Latent learning is learning that has occurred but is not translated into behavior. A scenario of latent learning would be if someone does not know something but wants to learn it. They eventually try it, and later on learn and know it. In this experiment, Tolman used rats as subject in order to see if reinforcement plays a role in the way that rats learn. The result of this experiment supported Tolman's hypothesis which was that learning remained latent until the organism had a reason to use it. Tolman also conducted experiments involving latent extinction. Latent extinction was the discovery that animals that have experienced a goal in the past, no longer need to obtain reinforcement to terminate a previously learned response to that goal much faster than animals without experience. For example, if someone used to live in a particular location but moved and came back to their location a year later, they will still be familiar with that location from memory, compared to someone who is new to that location.

B.F. Skinner was another significant behaviorist theorist. Skinner was a radical behaviorist and refused to acknowledge any part of mental events in human behavior. Functional analysis is a study focused on the relationship between reinforcements and response rate or probability. Skinner used this approach when describing environmental and behavioral events. Skinner studied Thorndike's experiments, and he interpreted them into his own ideas which led to his theory of operant behavior. Operant behavior is produced by an organism instead of being caused by a known stimulus. For example, there is a ¿½cry out¿½ method in which a parent does not respond to a crying child. As a result, the child learns that their crying does not elicit a parent's response and is insignificant. Reinforcement and punishment are also significant to operant behavior. To Skinner, reinforcement is anything that increases the rate or probability of a response. Change in an organism's behavior is identified through the use of reinforcement. He discovered that punishment does not weaken behavior, but strengthens behavior.

Skinnerian principles were developed due to differences that Skinner approached in comparison to other behaviorists. He applied one general rule, that if reinforcement contingencies are changed, then behavior will change. Behavior therapy was another Skinnerian principle that uses learning to treat emotional and behavioral disorders such as drug addiction, mental retardation or speech disorders. Token economies were also developed through the Skinnerian principles. Desirable behavior is strengthened by using valuable tokens as reinforcement. An example of these tokens can include candy, cigarettes, or coffee. Skinner's principles played a huge role in the development of behavior modification procedures which helped improve the quality of life for many people.

Psychologists today refer to behaviorism as part of cognitive psychology. Cognitive psychology is a branch of psychology that studies mental processes such as how people think, perceive, remember, and learn, but modern cognitive psychologist continue to use behavior indexes. Cognitive psychology is associated with all aspects of behaviorism, except for Tolman's views. Psychologists continue to base psychology on overt behavior. Many psychologists who categorize themselves as experimental psychologists are behaviorists.

Behaviorism remains the basis of all fields of psychology. Like other fields in psychology, behaviorism has its roots in philosophy, and continues to develop from the original theorists, to the new developments of today. The three main theorists John B. Watson, Edward Chace Tolman, and B.F. Skinner are vital not only to behaviorism, but psychology as a whole. The study of behaviorism will continue to progress as research and technology continues to develop. Without behaviorism, psychology would not be where it is today.

The study of human behavior is diverse and many studies have been done trying to find more about human behavior. Many researchers have concluded that human behavior is complex and sometimes unpredictable. The environment is one of the major factors in the development of human behavior. The unpredictability of human behavior tries to show that people are unique and operate on different ideas and backgrounds. In addition, the brain of a human being is independent to that of another person, therefore, this creates a diversity of how people think, act and make decisions. The study of human behavior requires a deep and thorough research on people in different places and environments. B.F. Skinner, a renowned researcher and psychologist, is one of the persons who engaged all his life studying human behavior and personality. He came up with various theories that help in understanding human behavior and determine their thinking and actions.

Human beings are unique in most of their operations. This is what makes the study of human is complex. This forces the researchers to have a thorough observation and deep study in their actions. This context will discuss the life of B.F. Skinner and his theories on human behavior. It will mainly focus on the contributions that Skinner has made in the field of psychology. In addition, it will explain some of the theories that Skinner discovered.

Burrhus Frederic (B.F.) Skinner was born on 20 March of the year 1904 in Pennsylvania, on a small town called Susquehanna. His parents were William and Grace Skinner. His father, William Skinner, was a lawyer. B.F. Skinner turned to be an atheist when a Christian teacher tried to convince him about hell, which he feared very much from the description given to him by his grandmother. Skinner had a brother named Edward Skinner, who died aged sixteen years due to cerebral hemorrhage. B.F. Skinner described his life in Susquehanna, Pennsylvania as stable and warm. He died in August 1990, suffering from Cerebral Cancer. In his childhood, he could be seen to enjoying constructing and inventing things in his childhood. This was a good base for his profession for applying in psychological experiments.

B.F. Skinner went to Hamilton College in New York, where he obtained a bachelor in Administration in English literature in 1926. Skinner linked to the Lambda Chi Alpha while still in Hamilton College. He was involved in writing school papers. Due to his atheist nature, he was desperate of the school he attended, which was religious. After college, he spent his time struggling to write about fictional ideas. Later, he became disappointed by his writing skills and made a conclusion that he possessed little experience of the world and hence he had no strong impression on what to write. After this conclusion, he discovered the writings by Pavlov and John Watson. These writings made him to develop a good basis of his research and opened a new way of understanding the natural things and behaviorism. John Watson, a writer, influenced Skinner with his writings about behaviorism.

The inspiration by John Watson made Skinner to have the desire to advance his study in psychology. This was followed by his enrollment in Harvard University for a graduate program in Psychology at the age of twenty-four. In Harvard, he met William Crozier, a researcher in the physiology department. Crozier was an advocate for animal study and measures in behavior. Using the functional and behaviorist background from William, Skinner started to his studies on behaviorism. He combined theories and methods from physiology and psychology where he developed various ways of data recording and analyzing. This was facilitated by the diversity of working across two interrelated disciplines. He used rats as his specimen for experiments. Using his new methods, he noticed that his observations were affected influenced by the preceding and subsequent responses. He focused on what prevailed after a particular behavior, which made him note that the effects of an action can influence the e learning of an organism.

Skinner realized that the behavioral approaches that were used at that time depended on the Watson and Pavlov's works. These two behavioral works were focused on the paradigm of Stimulus- response and their method of conditioning aimed at finding out the response and effect of those stimuli on learning of organisms. Therefore, he decided to focus on another method of approaching behavioral response and their effect learning.

Skinner received his PHD of psychology in Harvard in 1931 and continued with his study until 1936. In the same year, he married Yvonne Blue. Later, he taught in Minnesota University and University of Indiana. In Indiana University, he acted as a chairperson of the psychology department from 1946 to 1947. His service as a chairperson gave him the opportunity to advance his knowledge in human behaviorism. Working as the head of a whole department, provides a wide range of experience such as management skills and learn how to deal with humans. It is in this service at the university where he published his first book titled Verbal Behavior. In his book, he tried to explain the development of language in human beings. The element of language was a factor that affected human behavior, hence his book. In 1958, he returned to Harvard University as a permanent Professor and remained there for the remaining part of his life.

In his theories, Skinner was dedicated in developing his particular brand called Radical behaviorism. He defined Radical behaviorism as the philosophy of science of behavior, which strives to realize behavior as dependent on environmental histories for consequence reinforcement. Skinner started working on the issues of human behavior. He believed that it was a waste of time trying to study the hidden and unconscious aspects of humans. This belief made him to reject various theories, which were prominent in the psychological discipline (Kuhlman, 2005). He believes that the core of behaviorism lies on learning and reinforcement of human and animal behavior.

Skinner focused his time in studying observable kinds of human behaviors. He spent much of his career trying to refine the theories of reinforcement. He always said that personality is an element that develops in human beings and the behavior responds depending on the external events. He defined personality as an aspect that consists of the collection of human behaviors in a certain environment. He proposed that the differences in learning experiences are the major reasons behind the difference in human behavior. Therefore, learning differences affects human personality. This means that people are the way they are because they are rewarded to be that way. Skinner stated that this belief disregards thoughts, emotions and choice of human freedom (Kuhlman, 2005).

He continued to say that it is natural for human beings to behave in certain ways so that one can get favorable rewards. This means that if a person wants to experience reinforcement, the person develops personality traits, which are positive. Some of the traits acquired are included in category of agreeableness of the Big Five, which include understanding, empathetic, compassionate and positive thinker. He added that for a person to change his/her negative traits, the environment must be changed before any other element. This behaviorist viewpoint is strict and it tries to oppose the beliefs of other psychologists. This viewpoint requires that a person must change the inner self before change can be fully experienced. The inner self is the person's personality traits.

According to Skinner, he says that the change of behavior of humans aims at improving or worsening their situation. The behavior change brought about by environment affects the body condition, which results in some rewards (Kuhlman, 2005).As Skinner puts it; behavior is learnt depending on the available conditions. A person may develop behaviors due to the certain surrounding, which affects their personality (Kuhlman, 2005). It was Skinner's believe that it was a waste of time trying to study the hidden and unconscious aspects of humans. This belief made him to reject various theories, which were prominent in the psychological discipline. Skinner's believes were his inspirations, which made him remain strong regardless of the critics and opposition.

Until today, Skinner is famous for his theory of human behaviorist. Other psychologists believed that human psychology is a complex system due to its varied influences, but Skinner had believe that human psyche is not really varied to the cognitive processes of other animals. He stood on the opposition of human psychology in his life career where he denied the aspect of humans possessing freedom and dignity. Skinners study and theories were based on self-observation. This made him to be a supporter for behaviorism, where he stood onto his believes. According to Skinners theory of Radical Behaviorism, it observes reinforcement as a central concept of behaviorism and as a central tool for controlling and shaping of behavior. A major misconception about this theory is that negative reinforcement is identical to punishment. Both positive and negative reinforcement promotes or enhances the probability of a certain behavior reoccurring. Through his study and research, Skinner found that the theory of behaviorism could be applied in natural selection in the broadest context of organism behavior.

In his work, Skinner designed some tools and equipment to use in his experiments and study. Some of the equipment includes cumulative recorder, teaching machine and the verbal summator. The cumulative recorder and teaching machine were used mainly in the process of conditioning. Conditioning process is the method that skinner used to control the behavior of an organisms by managing the rewards and punishments in the surrounding. Cumulative recorder was used to record behavior automatically and graphically. On the other hand, the teaching machine was deployed to administer the course of programed instructions. Skinner used these own inventions in his study of behavior and human psychology Other Skinners inventions include the operant conditioning chamber, air crib and the pigeon guided missile. The US navy used the pigeon-guided missile in controlling missile in their battle against the Germans. This battle invention depended on human's behavior of ignorance. This is because no human could detect in time the pigeon as guiding a missile.

By considering the process of conditioning, Skinner used two types of conditioning, which includes classical and operant conditioning. Classical conditioning involves pairing a neutral stimulus with an unconditioned stimulus so that it is capable of bringing a previously unconditioned response, which is called the conditioned response. On the other hand, operant conditioning involves use reinforcement to enhance the probability that a certain behavior will be repeated. Skinner used the teaching machine and the cumulative recorder devices to implement the operant conditioning process.

As seen in this context, Skinner has shown to various contributions to the discipline of psychology. He participated in publishing more than two hundred articles and over twenty books in psychology. Due to his authorship, he was largely praised as the most instrumental and influential psychologist in the 20th century. The work he performed in operant conditioning remains crucial in teaching in the present day. This is because professionals in mental health use the operant conditioning while dealing with clients. In addition, teachers use punishment and reinforcement to shape behavior in class setting. Furthermore, animal trainers use Skinner's techniques to train dogs and other animals. From these contributions of Skinner, he scooped various awards such as the National Medal of Science, Human of the Year Award, Gold Medal of the American Psychological Foundation and Citation for Outstanding Lifetime Contribution to Psychology. Some of his Psychological publications include Superstition in the Pigeon, Two Types of Conditioned Reflex and Pseudo Type and The Origins of Cognitive Thought.

From this context, it can be seen that B.F. Skinner is a major contributor in the development and growth of psychology. His theories have been used in various fields, which involve dealing with human- beings. In this world, many fields involves dealing with humans, therefore, Skinner's theories and inventions are used in most fields such as health, scientific research and security. From Skinner's beliefs, people can learn that behaviorism is an aspect, which affects the human personality depending on the environment one is living. One can also learn that change of human behavior can only happen by changing the environment. Much can be learnt from Skinner's psychological career. His ideologies are based on deep study and research through experiments, which used own inventions for the best results. Skinner's inventions and theories will live to be remembered for many years.

EARLY INFLUENCES

Prehistoric Period

Cave painting of a horse at Lascaux Our early ancestors critically depended on a detailed understanding of animal behavior. Specifically, knowledge of habitat preference, movement patterns, and sensory biology of prey as well as predators were essential. Rock paintings, figurines, and carvings, depicting scenes from daily life and religious rituals, commonly feature detailed likenesses of many kinds of animals. Early forms of rock art emerged during the Paleolithic (200,000 BCE), however, the advent of modern man brought about extraordinary cave wall paintings (40,000 BCE) of animals, hunting scenes and other graphic illustrations of prehistoric life. Holes in Neolithic skulls illustrate that patients dating back to such early times have been subjected to head surgery, presumably with the aim of curing headaches or mental disorders. Sumerian descriptions show the use of Opium, derived from poppy plants, to produce euphoriant effects.

Native American

A common theme contends that the world was populated by many peoples who were subsequently transformed into different kinds of animals. Thus all animals and humans share the presence of a soul. The close bond between them is shown in frequent rituals where the behavior of different animals is simulated. Imagery of animals is common. The Great Serpent Mound represents an effigy mound located on a plateau along Ohio Brush Creek in Adams County, Ohio. Conforming to the topography, the snake's head approaches a cliff point, the snake then winds back and forth for more than 200 meters and ends with a triple coiled tail.

Egyptian Mythology

Horus, an ancient Egyptian falcon headed-deity. Ancient Egyptian societies showed a keen interest in the temperaments and sensory abilities of different animals and ascribed them to individual deities.

Ancient Egyptians had a highly developed view of the afterlife with elaborate rituals for preparing the body and soul for a peaceful existence after death. Beliefs about the soul and its needs led to numerous ceremonies for the preservation of the body. The brain apparently was not considered an essential element for eternity, as it was treated with a distinct lack of dignity. It was often spooned into small pieces and removed through the nose, then unceremoniously discarded.

Classical Antiquity

Ancient Greek Period

Plato held the notion that the human mind as an entity pre-existed somewhere in the heavens, before being sent down to join a body on earth. Hippocrates believed that the brain was not only involved with sensation but was also the seat of intelligence. Like many before and after him, Aristotle was intrigued by the natural world around him. He, however, stands out as a seminal thinker on a wide range of subjects as he was able to move beyond his substantial abilities at observation by forwarding daring hypotheses in order to categorize and make sense of what he saw. Aristotle provided the first written records of mutualistic ties between individuals, animal tool use, and brood parasitism. His views on human nature considered our behavior as constituting formed habits and a desire for happiness. With his proposed concept of a tabula rasa he forwarded the thesis that individuals are born with no innate or built-in mental content, in a word "blank", and that their entire resource of knowledge is built up gradually from their experiences and sensory perceptions of the outside world. Herophilus studied the nervous system and distinguished between sensory nerves and motor nerves. Erasistratus studied the brain and distinguished between the cerebrum and cerebellum.

Roman Empire

As a physician in a gladiator school, Galen gained a unique view of the behavioral deficits that resulted from acute head trauma. Not surprisingly he wrote extensively on the connection between paralysis and severance of the spinal cord. His many audacious operations, including brain and eye surgery, were not again used for almost two millennia.

Middle Ages

Naturalists and philosophers focused on classical texts and, in particular, on the Bible in Latin. Accepted were Aristotle's views on all matters of science, including psychology. As knowledge of Greek declined, the West found itself cut off from its Greek philosophical and scientific roots. General levels of literacy further decreased as learning became largely restricted to monastic and cathedral schools that were primarily aimed at training clergy. The Church exerted an overpowering role in philosophical thought and nontheological scholarship was rare. Despite this many influences on the study of behavior in later periods began to emerge.

Arabists

Arabian scholars describe cranial and spinal nerves, along with surgical procedures for neurological disorders.

RENAISSANCE

The principal features of the Renaissance (14th-16th century) were the revival of learning based on classical sources and the broad advancements of science. It pioneered scientific study through deductive reasoning. René Descartes supplied a foundation for the pursuit of knowledge.

- Unity of all sciences through deductive reasoning
- Never accept anything as true that is not known clearly to be such
- Divide problems into as many parts as possible, and proceed from the simplest to the hardest

He argues that he perceives his body through the use of the senses, which often proven unreliable. So he concludes that thinking is his essence as it is the only thing about him that cannot be doubted. Descartes defines "thought" as every activity of which a person is immediately conscious. In his Mind/body dualism he viewed the body as a machine that follows the laws of physics. In contrast, he regarded the mind (or soul) as a nonmaterial entity that is not subject to such physical constraints. The Cartesians (i.e., the followers of Descartes) aimed to explain behavior using simple, natural, and entirely mechanistic principles. In their view only humans were endowed with an ability to reason.

With the advent of the scientific revolution, medieval worldviews steadily gave way to increasingly analytical methodologies and a weakening of vitalist traditions. Although medical thought continued to be guided by the occult, critical thinking began to chip away at the long-held notions based on Galen. Hippocrates' views are revived and Renaissance artists such as Leonardo da Vinci conducted detailed studies of human anatomy in order to better portray the human form. As Andreas Vesalius corrects Greek medical errors, he revolutionizes medicine. William Harvey applies scientific methodology to human physiology and solves the human circulatory system. Paracelsus rejects occultism and pioneers the use of chemicals and minerals in medicine.

Enlightenment

Thinkers of the Age of Reason (18th century) perceive their efforts as bringing to an end previous periods of irrationality, superstition, and tyranny. John Locke's philosophy revives Aristiotles theory of a tabula rasa, where the (human) mind is at birth a "blank slate" without rules for processing data, and where data is added and rules for processing are formed solely by one's sensory experiences. The concept of a blank mind at birth emphasized the individual's freedom to author one's own human character. Moreover, as members of ones society will become what society make of them, it becomes essential to assure universal access to learning opportunities, medical care, and possibilities for personal development.

Evolution

In the second half of the 19th century the simple taxonomic classification of all life forms was supplanted by a growing focus on parsimonious and scientifically testable (phylegenetic) hypotheses that could explain the relatedness among groups of organisms. Studies of the fossil record, the geographic distribution of faunas, and the characteristics that linked the diversity of live, convinced a growing number of scientists that species indeed changed over time. Evolution became viewed as a branching process, whereby populations are altered over time and may specialize into separate branches, hybridize together, or terminate by extinction. However, the mechanism driving these changes remained unclear until the 1859 publication of Charles Darwin's On the Origin of Species, which provided the theory of evolution with the likely mechanism of natural selection.

- Charles Darwin (12 February, 1809 - 19 April, 1882) represents arguably among the most significant scientists of all time. He realized and presented compelling evidence for natural selection as the main mechanism by which all species of life have evolved over time from common ancestors. Evolution, as an explanation for the diversity of life, had emerged earlier and was generally accepted by most in his lifetime. His 1859 book On the Origin of Species established evolutionary descent with modification as the dominant scientific explanation of diversification in nature and this view emerged as the primary explanation in modern evolutionary theory. He also examined human evolution and sexual selection in The Descent of Man, and Selection in Relation to Sex, followed by The Expression of the Emotions in Man and Animals. Darwin suggested that all human characteristics are fundamentally inherited and all have been subject to evolution. Emotions may best be understood as consequences of adaptive reactions.
- George John Romanes, a Canadian-born English evolutionary biologist and physiologist, laid the foundation for comparative psychology, postulating a similarity of cognitive processes and mechanisms between humans and animals.

PSYCHOPHYSICS

Psychophysicists, such as Gustav Theodor Fechner and Wilhelm Wundt, aimed to relate matter to mind via a study of inner experiences or feelings towards defined physical stimuli. Psychophysicists explicitly rejected the concepts of vitalists, i.e. that the functions of a living organism are due to a vital principle distinct from physicochemical forces. The main focus lay on the way subjects perceive experimental stimuli that can be objectively measured, such as pure tones varying in intensity, or lights varying in luminance. A broad range of senses has been studied, including vision, hearing, touch (including skin and enteric perception), taste, smell, or the sense of time.

Psychoanalsis

- Psychoanalysis is a family of psychological theories and methods based on the pioneering work of Sigmund Freud. As a result of talking with his patients, he came to believe that psychopathologies stemmed from culturally unacceptable, thus repressed and unconscious, desires and fantasies of a sexual nature. Moreover he stressed the presumed value of dreams as sources of insight into unconscious desires. The analyst's goal is thus aimed to discover connections among the unconscious components of patients' mental processes. This ought to help liberate the patient from unexamined or unconscious barriers of transference and resistance, that is, past patterns of relating that are no longer serviceable or that inhibit freedom.

- Carl G. Jung's unique and broadly influential approach to psychology emphasized understanding the psyche through exploring the worlds of dreams, art, mythology, world religion and philosophy. Jung emphasized the importance of balance and harmony. He cautioned that modern humans rely too heavily on science and logic and would benefit from integrating spirituality and appreciation of the unconscious realm. His Collective unconscious refers to a human experience reservoir, a kind of knowledge, we are all born with but can never be directly conscious of. His Archetypesrepresent generic, idealized models of personalities with unlearned tendencies to experience things in a certain way.

Comparative Psychology and Behaviorism

This school of thought represents a movement in psychology that advocates the use of strict experimental procedures to study observable behaviors (or responses) in relation to environment (or stimuli). Behaviorists regard learning and environmental conditions as dominant over the effects of genetics and heredity. Concepts, such as learning or memory are explored using artificial tasks in strictly controlled environments (e.g., Skinner Box, Morris' water maze, Radial arm maze).

Comparative psychology viewed science less as a framework for explaining events but rather for predicting and controlling them. Adoption of the observational techniques that had proven useful in the study of the behavior of animals was viewed as the best way to advance the scientific study of human beings. The external environment represented the main determining influence on behavior and its manipulation held the most promise for improving the condition of human beings. The concept of a tabula rasa was revived.

A common tenet held that the complex phenomena of behavior could be explained by and reduced to clear, distinct, reflexive units of behavior.

William James described psychology as "the science of mental life." As a rigid application of the scientific method is only possible for learned behaviors, he felt that our studies should foremost focus on studying those

Ivan P. Pavlov realized that behavior changes when an organism comes to associate one stimulus with another. Specifically, a reflexive or automatic response transfers from one stimulus to another - Classical Conditioning.

Edward L. Thorndike Using a Puzzle Box, he demonstrated that behaviors that are followed by a positive outcome are often repeated, while those followed by a negative outcome or none at all are extinguished (i.e., Operant Conditioning). He postulated the Law of effect where behavior changes when the results of an action proved of consequence (e.g. rewarding, noxious).

John B. Watson stimulus-response theory of psychology, emotional reactions are learned in much the same way as other skills

B.F. Skinner contends that nearly all behavior is shaped by complex patterns of reinforcement in a person's environment. Skinner Box: An animal placed inside the box is rewarded with a small bit of food each time it makes the desired response, such as pressing a lever or pecking a key. A device outside the box records the animal's responses.

Ethology

"When, as by a miracle, the lovely butterfly bursts from the chrysalis full-winged and perfect,... it has, for the most part, nothing to learn, because its little life flows from its organization like melody from a music box" -- Douglas A. Spalding, 1873

Ethology, the biological study of behavior under naturalistic conditions, emerged out of diverse roots by individuals largely trained in a zoological tradition. It aimed to provide a more comprehensive study of animal behavior than previously available. Where comparative psychology studies the behavior of animals in the context of what is known about human psychology, ethology explores animal behavior in the context of animal anatomy and physiology. Ethologists specifically considered the significance of inherited elements of behavior alongside learned components. Influenced by evolutionary biologists such as Thomas Hunt Morgan and others, many aspects of behavior may indeed be accessible to selection. Charles Otis Whitman examined the importance of phylogeny to the origins of behavior. Oskar Heinroth rather focused on how the adaptive value of behavior may explain behavioral differences among taxa. Experiments, in which he raised young birds separate from other individuals of their own species, demonstrated that many species-specific behaviors were nonetheless present. Interest in genetically-programmed elements (instincts) were contrasted to those subject to learning (e.g., imprinting). Using a comparative approach ethologists are thus interested to search for the biological basis of behavior (including human behavior) by comparing specific activities to those of close relatives.

Karl von Frisch studied the senses of honey bees, identified their mechanisms of communication, and showed their sensitivity to ultraviolet and polarized light. He unlocked the basic secrets of the forager's waggle dance, research which continues to yield new insights to this day. Techniques developed by behaviorists, such as classical and operant conditioning, have proven valuable to ethologists. Animals trained to respond to a particular stimulus in the appropriate behavioral context can demonstrate their ability to discriminate between different types of stimulus. For example, von Frisch used classical conditioning to show that fish could hear, and bees were able to perceive colors.

Konrad Lorenz' work included studies of imprinting (i.e., the capacity to learn specific types of information at certain critical periods in development) and instinct (i.e., the capacity of an animal to complete a complex behavior in response to a first-time encounter with a given situation). Filial imprinting refers to the case where young of a species need to acquire the image of what their parents look like (e.g. ducks). Instinctive behavior can usually be elicited by a limited combination of complex environmental cues. Such a sign stimulus refers to a particular subset of features or qualities of a natural stimulus or from complex environmental cues that is sufficient to elicit a specific behavioral response. They noted that different individuals produce identical behavioral response to specific sign stimuli - Motor patterns (or Fixed Action Pattern, FAP). Once initiated, FAPs continue until completion. A releasing mechanism is a functionally organized, neural circuit that recognizes a specific releaser and produces the appropriate response. Due to their stereotype, behaviors can be cataloged in form of an Ethogram (i.e., a comprehensive compilation of the behavioral repertory of a species, a catalog of actions as complete and precise as possible). An internal factor that determines how likely an animal is to perform a behavior (i.e., Action-Specific Energy).

The causation of behavior is viewed in the context of a metaphor - the Psychohydraulic model. Action specific energy is akin to a liquid building up in a vessel. A high level of action specific energy results when a behavior has not been performed in some time. Pressure from the water pushes onto a piston which is also pulled by sign stimuli in the environment, represented by weights. As pressure mounts the piston finally opens a valve that allows water to flow out (representing the performance of the behavior) Example: An animal that has had a good meal will have a low drive to eat more (i.e., little water in the apparatus). As it goes for a while without eating, the drive (i.e., hunger) builds up. If food is then encountered, the animal will try to consume it. If no good food is encountered, less preferred items will be accepted. This model has faced much criticism as performance of many behaviors appear to increase the tendency to perform it again (i.e aggression), rather than bring about a decrease.

Three Ethologists shared the 1973 Nobel Price for Physiology or Medicine in recognition of their work over the previous 30 years as founders of the field of ethology.

- Karl von Frisch social behavior of bees
- Konrad Z. Lorenz studied the behavior of geese, ravens, and ducks

- Niko Tinbergen behavioral studies of gulls and stickle backs. "Man likes to consider himself a rationale being; he not only fails to recognize his instincts, he even scoffs at instincts and considers them inferior. This attitude makes him fall victim to similarity between man and animals; he recognizes himself only too often in an animal. Also, he has rather a higher opinion of instincts; for instance, he recognizes that they are at the bottom of his subjective experiences. And he has doubts about the all-controlling power of reason, for he sees its limits too often."

BEHAVIORAL GENETICS

Behavioral Genetics explores the hereditability of behavioral traits. More recently, this has given rise to a largely quantitative emphasis on using molecular genetic techniques in order to isolate individual genes that play a role in behavior, neural and behavioral pathologies, or individual human characteristics.

Each inherited feature of an organism is determined by a section of genetic code - a gene which represents the smallest hereditary unit. The genome refers to one individual's entire set of hereditary information and includes genes as well as non-coding sequences of DNA. The genomes for any two individuals, except for those of identical twins or cloned animals, are unique. Such differences (i.e., polymorphism) produce different versions of the same genes, called alleles. Despite these differences in genetic makeup, we also find major commonalities across much of the code shared by all forms of life. The specific genetic makeup of an individual is called its genotype. A trait refers to a specific characteristic or feature of an organism (although in psychology the term may be used to represent a component of personality). The term phenotype represents the specific manifestation of a trait or its physical appearance and constitution and refers to the actual state of that trait (e.g., the trait eye color has the phenotypes blue, brown and hazel). The individual's phenotype emerges as interaction of genetic information with environmental conditions.

Sir Francis Galton (February 16, 1822 - January 17, 1911), a younger relative of Charles Darwin, was an English explorer, geographer, and statistician. Intrigued by Darwin's concept of natural selection and with a keen interest in the breeding of domesticated animals, Galton explored variation in human populations and its implications. He became interested at first in the question of whether human ability was hereditary. He presented evidence for the inheritance of a range of abilities - "a man's natural abilities are derived by inheritance, under exactly the same limitations as are the form and physical features of the whole organic world". He also proposed adoption studies, including trans-racial adoption studies, to separate out the effects of heredity and environment. He was instrumental in the development of quantitative methods to test human intelligence, statistical correlation, fingerprint analysis. self-report questionnaires, and word-association studies. In a darker side, he became a vociferous proponent of eugenics where he proposed to improve the human race via selective breeding of those individuals who managed to score high marks on intelligence tests.

- John Paul Scott - genetics of behavior by studying personality traits in breeds of dogs

Behavioral Ecology

"Nothing in biology makes sense except in the light of evolution..." -- Theodosius Dobzhansky Behavioral ecology is the study of the ecological and evolutionary basis for animal behavior. Primarily concerned with ultimate (evolutionary) consequences, researchers aim to identify phylogenetic contingencies within behavior (i.e., to what degree behavior is inherited from its ancestors) and to examine its functional significance (i.e., how the behavior impacts the bearers reproductive success). Behavioral ecologists contend that in order to minimize ad-hoc explanations arising from anecdotal observations a rigorous examination of the adaptive consequences of a behavior are essential. Specifically, it explores what a trait is good for in an evolutionary context.

Optimization theory is concerned with assessing the success of behavior. It attempts to identify those behavioral strategies that offer the highest return under a given set of conditions using a cost/benefit analysis. In this context success or fitness is judged by considering the number of offspring that the bearer of the behavior would contribute to the next generation relative to its suboptimal conspecifics. Optimization theory, such as optimal foraging theory, states that individuals ought to adopt strategies that allow it to consume the most energy while expending the least amount of energy. Evolutionary game theory asserts that the success of strategies should not only consider its inherent benefits, as the actual success depends on what other strategies exist in the population. An evolutionarily stable strategy is a strategy which when adopted by all members cannot be beaten out by any invading, mutant strategy.

- John R. Krebs - acoustic communication in birds and crickets, optimal foraging theory
- N. B. Davies - the economics of behavioral strategies

Sociobiology

Sociobiology aims to explain behavior in all species by considering the evolutionary advantages of social behaviours. Sociobiology is based on the idea that all animals, including humans, will tend to act in ways to improve their own reproductive success, and that this will result in social processes conducive to genetic fitness. For example, cultural practices and values related to sexuality are viewed as shaped by natural selection (e.g., societal rules for incest avoidance). The concept has been criticized for its reliance on innate predispositions over learning, and by the view that presents situations are not representative of historical conditions.

The selfish genes hypothesis states that the individual is not as important to the population as its genes. Kin selection refers to changes in gene frequency across generations when biological relatives influence the reproductive success of each other. Thus, a gene that prompts behavior which enhances the fitness of relatives but lowers that of the individual displaying the behavior, may nonetheless increase in frequency, as relatives often share the same genes. The enhanced fitness of relatives can at times more than compensate for the fitness loss incurred by the individuals displaying the behavior.

- E.O. Wilson American entomologist, naturalist, and environmentalist.
- William D. Hamilton British evolutionary biologist, considered one of the greatest evolutionary theorists of the 20th century. Hamilton became famous for his theoretical work on kin selection - Hamilton's rule.

Evolutionary Psychology

Evolutionary Psychology attempts to explain "useful" mental and psychological traits-such as memory, perception, or language-as adaptations, i.e., as the functional products of natural selection.

Neurosience

The last frontier of the biological sciences--their ultimate challenge--is to understand the biological basis of consciousness and the mental processes by which we perceive, act, learn, and remember. -- Eric Kandel, 2000

In 1791, Luigi Galvani discovered that the muscles of dead frog's legs twitched when struck by an electric spark. According to popular version of the story, Galvani had been conducting experiments with static electricity by rubbing frog skin. Galvani's assistant touched an exposed sciatic nerve of the skinned frog carcass with a metal scalpel that had picked up a charge, they saw sparks and the dead frog's leg kicked as if in life. The observation made Galvani appreciate the relationship between electricity and muscle movement. Galvani's cousin. who engaged in a public demonstration of electrical stimulation of muscles on the body of an executed criminal, reputedly provided the inspiration for Mary Shelley's Frankenstein. With stimulation of the body's face, the jaws began to quiver, facial muscles were horribly contorted, an eye opened, the right hand was raised and clenched, and muscles in legs and thighs moved.

The late 19th century witnessed the rise of modern neuroscience. Hypotheses that distinct brain functions were contained within specific regions of the brain emerged with evidence from patients with epileptic seizures, head trauma, and stroke. Extensive characterizations of neurons throughout the brain led Camillo Golgi and Santiago Ramón y Cajal to propose the neuron doctrine, i.e., that the function of the brain is a product of the electrical and chemical activity of individual neurons. Further support came from the discovery that muscles and neurons showed electrical excitability, which could be transferred onto neighboring cells. Electrical properties are assessed either by recording from an electrode placed into the lumen of a cell (intracellular) or from its surrounding vicinity (extracellular). Intracellular recording techniques aim to either measure the current needed to maintain a constant voltage (voltage clamp) or measure the voltage needed to keep the current constant (current clamp) across the cell membrane. Using such techniques, Alan Hodgkin and Andrew Huxley identified the ionic mechanisms that underly the electrical properties of neurons in giant axons of squid. John Carew Eccles identified synapses as the sites of communication between neurons.

Generations of neuroscientists had perceived the brain as a rather passive, reactive organ whose functional capacities were maintained by a sustained flow of sensory input.

NEUROETHOLOGY AND BEHAVIORAL NEUROSCIENCE

"It is an urgent task of ethologists and neurophysiologists to join efforts in the training of 'etho-physiologists'" -- Tinbergen, 1951. The Study of Instinct

Neuroethology is a synthesis of neuroscience and the study of behavior that explores neural mechanisms of natural behavior. This is in contrast to other approaches to neuroscience that study the nervous system in isolation, or in the context of artificial conditions. The neuroethological approach stems from the idea that nervous systems have evolved to address problems of sensing and acting in certain environmental conditions. The functions of nervous systems, therefore, are best understood in the context of the problems they have evolved to solve.

Using physiological experiments, Erich von Holst showed that behaviors can be elicited experimentally even in the absence of any sensory input. In contrast to Sherrington's reflex theory, van Holst postulated that basic central nervous function required permanent production of endogenous inhibitory stimulation. Behavior emerged when inhibitory control was diminished. He also demonstrated that internal, autonomous, rhythmic behavior can be independent of environmental factors. By slicing a worm into separate segments, and attaching each segment to a sensitive voltmeter, he noticed distinct, consecutive deflections on the meter which demonstrated as a wave moving through the severed parts from the front to the end of the entire cut-up specimen at approximately the speed of a contraction wave of a wriggling earthworm.

Behavioral Neuroscience, a subdicipline of neuroscience, is largely concerned with the function of neural systems in generating behavior.

Mechanisms of sensory processing have utilized a variety of model systems, including the moth's acoustic startle response to bat sounds, jamming avoidance in electric fish, or sound localization in owls. A detailed understanding of the control and coordination of movement patters has emerged from investigations of food processing in lobsters, prey capture in frogs, or swimming in lampreys. Advances in our knowledge of the hormonal control of behavior have profited from work on the division of labor in bee colonies, behavioral changes due to molting in moths, and aggression in lizzards.

Sensory receptors and neurons exhibit functional tuning to a highly specific subset of stimulus conditions. For instance mechanoreceptors with sensory hairs detect air currents in crickets by transducing both direction and velocity of relevant stimulus energies. Directional tuning is determined by the preferred orientation of the hair's socket. Velocity tuning occurs when longer hairs respond to low wind velocities while shorter hairs need higher wind velocity to activate them. Each mechanoreceptor is innervated by a single sensory neuron and its directional tuning curve represents the cell's sensitivity to wind direction. Most substrates for sensory processing show a strict spatial organization - its topographic mapping. A system for extracting temporal information can be constructed via delay lines, a system which utilizes the fact that electrical conduction in neurons is not instantaneous; it is delayed depending on the length of the axon and other properties of the neuron (e.g. myelination). The resulting delays can be used for time sensitive calculations, such as calculation of inter-aural time differences in sound localization. A central pattern generator (CPG) is a network of neurons (or even a single neuron) which is able to exhibit rhythmic activity in the absence of sensory input.

With respect to causation, the question becomes: What makes the behaviour happen? To answer this question, it becomes important to identify the physiological and cognitive mechanisms that underlie the tail-wagging behaviour. For example, the way the dog's hormonal system adjusts its responsiveness to stimuli, how the dog's nervous system transmits signals from its brain to its tail, and how the dog's skeletal-muscular system generates tail movements need to be understood. Causation can also be addressed from the perspective of cognitive processes (that is, knowing how the dog processes information when greeting a companion with tail wagging). This perspective includes determining how the dog senses the approach of another individual, how it recognizes that individual as a friend, and how it decides to wag its tail. The dog's possible intentions (for example, receiving a pat on the head), feelings, and awareness of self become the focus of the investigation.

With respect to ontogeny, the question becomes: How does the dog's tail-wagging behaviour develop? The focus here is on investigating the underlying developmental mechanisms that lead to the occurrence of the behaviour. The answer derives from understanding how the sensory-motor mechanisms producing the behaviour are shaped as the dog matures from a puppy into a functional adult animal. Both internal and external factors can shape the behavioral machinery, so understanding the development of the dog's tail-wagging behaviour requires investigating the influence of the dog's genes and its experiences.

With respect to function: How does the dog's tail-wagging behaviour contribute to genetic success? The focus of this question is rooted in the subfield called behavioral ecology; the answer requires investigating the effects of tail wagging on the dog's survival and reproduction (that is, determining how the tail-wagging behaviour helps the dog survive to adulthood, mate, and rear young in order to perpetuate its genes).

Connect with Britannica

Lastly, with respect to evolutionary history, the question becomes: How did tail-wagging behaviour evolve from its ancestral form to its present form? To address this question, scientists must hypothesize evolutionary antecedentbehaviours in ancestral species and attempt to reconstruct the sequence of events over evolutionary time that led from the origin of the trait to the one observed today. For example, an antecedent behaviour to tail wagging by dogs might be tail-raising and tail-vibrating behaviours in ancestral wolves. Perhaps when a prey animal was sighted, such behaviours were used to signal other pack members that a chase was about to begin.

Both the biological and the physical sciences seek explanations of natural phenomena in physicochemical terms. The biological sciences (which include the study of behaviour), however, have an extra dimension relative to the physical sciences. In biology, physicochemical explanations are addressed by Tinbergen's questions on causation and ontogeny, which taken together are known as "proximate" causes. The extra dimension of biology seeks explanations of biological phenomena in terms of function and evolutionary history, which together are known as "ultimate" causes. In biology, it is legitimate to ask questions concerning the use of this life process today (its function) and how it came to be over geologic time (its evolutionary history). More specifically, the words use and came to be are applied in special ways, namely "promoting genetic success" and "evolved by means of natural selection." In physics and chemistry, these types of questions are out of bounds. For example, questions concerning the use of the movements of a dog's tail are reasonable, whereas questions regarding the use of the movements of an ocean's tides are more metaphysical.

History and Background on Ivan Pavlov

Ivan Petrovich Pavlov was a Russian, and later Soviet, physiologist, psychologist, and physician. He was awarded the Nobel Prize in Physiology or Medicine in 1904 for research pertaining to the digestive system. Pavlov is widely known for first describing the phenomenon of classical conditioning.

In the 1890s, Pavlov was investigating the gastric function of dogs by externalizing a salivary gland so he could collect, measure, and analyze the saliva and what response it had to food under different conditions. He noticed that the dogs tended to salivate before food was actually delivered to their mouths, and set out to investigate this "psychic secretion", as he called it.

He decided that this was more interesting than the chemistry of saliva, and changed the focus of his research, carrying out a long series of experiments in which he manipulated the stimuli occurring before the presentation of food. He thereby established the basic laws for the establishment and extinction of what he called "conditional reflexes" - i.e., reflex responses, like salivation, that only occurred conditionally upon specific previous experiences of the animal. These experiments were carried out in the 1900s, and were known to western scientists through translations of individual accounts, but first became fully available in English in a book published in 1927.

Unlike many pre-revolutionary scientists, Pavlov was highly regarded by the Soviet government, and he was able to continue his research until he reached a considerable age. Moreover, he was praised by Lenin and as a Nobel laureate.

After the murder of Sergei Kirov in 1934, Pavlov wrote several letters to Molotov criticizing the mass persecutions which followed and asking for the reconsideration of cases pertaining to several people he knew personally.

In later life he was particularly interested in trying to use conditioning to establish an experimental model of the induction of neuroses. He died in Leningrad. His laboratory in Saint Petersburg has been carefully preserved as a museum.

Conscious until his very last moment, Pavlov asked one of his students to sit beside his bed and to record the circumstances of his dying. He wanted to create unique evidence of subjective experiences of this terminal phase of life.

The concept for which Pavlov is famous is the "conditioned reflex" (or in his own words the conditional reflex) he developed jointly with his assistant

Ivan Filippovitch Tolochinov in 1901. Tolochinov, whose own term for the phenomenon had been "reflex at a distance", communicated the results at the Congress of Natural Sciences in Helsinki in 1903. As Pavlov's work became known in the West, particularly through the writings of John B. Watson, the idea of "conditioning" as an automatic form of learning became a key concept in the developing specialism of comparative psychology, and the general approach to psychology that underlay it, behaviorism. The British philosopher Bertrand Russell was an enthusiastic advocate of the importance of Pavlov's work for philosophy of mind.

Pavlov's research on conditional reflexes greatly influenced not only science, but also popular culture. The phrase "Pavlov's dog" is often used to describe someone who merely reacts to a situation rather than using critical thinking. Pavlovian conditioning was a major theme in Aldous Huxley's dystopian novel, Brave New World, and also to a large degree in Thomas Pynchon's Gravity's Rainbow.

It is popularly believed that Pavlov always signaled the occurrence of food by ringing a bell. However, his writings record the use of a wide variety of stimuli, including electric shocks, whistles, metronomes, tuning forks, and a range of visual stimuli, in addition to ringing a bell. Catania cast doubt on whether Pavlov ever actually used a bell in his famous experiments. Littman tentatively attributed the popular imagery to Pavlov's contemporaries Vladimir Mikhailovich Bekhterev and John B. Watson, until Thomas found several references that unambiguously stated Pavlov did, indeed, use a bell.

Catania, A. Charles (1994); Query: Did Pavlov's Research Ring a Bell?, PSYCOLOQUY Newsletter, Tuesday, June 7, 1994 Behaviourism is primarily associated with Pavlov (classical conditioning) in Russia and with Thorndike, Watson and particularly Skinner in the United States (operant conditioning).

Behaviourism is dominated by the constraints of its (naÃ ve) attempts to emulate the physical sciences, which entails a refusal to speculate about what happens inside the organism. Anything which relaxes this requirement slips into the cognitive realm.

Much behaviourist experimentation is undertaken with animals and generalised. In educational settings, behaviourism implies the dominance of the teacher, as in behaviour modification programmes. It can, however, be applied to an understanding of unintended learning.

Classical condiioning is the process of reflex learning-investigated by Pavlov-through which an unconditioned stimulus (e.g. food) which produces an unconditioned response (salivation) is presented together with a conditioned stimulus (a bell), such that the salivation is eventually produced on the presentation of the conditioned stimulus alone, thus becoming a conditioned response. Such associations can be chained and generalised (for better of for worse): thus "smell of baking" associates with "kitchen at home in childhood" associates with "love and care". (Smell creates potent conditioning because of the way it is perceived by the brain.) But "sitting at a desk" associates with "classroom at school" and hence perhaps with "humiliation and failure"...

Behaviorism as a movement in psychology appeared in 1913 when John Broadus Watson published the classic article 'Psychology as the behaviorist views it'. Watson believed that all individual differences in behavior were due to different experiences of learning. Watson proposed that the process of classical conditioning (based on Pavlov's observations) was able to explain all aspects of human psychology. Everything from speech to emotional responses were simply patterns of stimulus and response. Watson denied completely the existence of the mind or consciousness.

Ivan Pavlov showed that classical conditioning applied to animals. Did it also apply to humans? In a famous (though ethically dubious) experiment Watson and Raynor (1920) showed that it did "Little Albert" was a 9-month-old infant who was tested on his reactions to various stimuli. He was shown a white rat, a rabbit, a monkey and various masks. Albert described as "on the whole stolid and unemotional" showed no fear of any of these stimuli. However what did startle him and cause him to be afraid was if a hammer was struck against a steel bar behind his head. The sudden loud noise would cause "little Albert to burst into tears.

When "Little Albert" was just over 11 months old the white rat was presented and seconds later the hammer was struck against the steel bar. This was done 7 times over the next 7 weeks and each time "little Albert" burst into tears. By now "little Albert only had to see the rat and he immediately showed every sign of fear. He would cry (whether or not the hammer was hit against the steel bar) and he would attempt to crawl away. Watson and Raynor had shown that classical conditioning could be used to create a phobia. A phobia is an irrational fear, i.e. a fear that is out of proportion to the danger. Over the next few weeks and months "Little Albert" was observed and 10 days after conditioning his fear of the rat was much less marked. This dying out of a learned response is called extinction. However even after a full month it was still evident

If, when an organism emits a behaviour (does something), the consequences of that behaviour are reinforcing, it is more likely to emit (do) it again. What counts as reinforcement, of course, is based on the evidence of the repeated behaviour, which makes the whole argument rather circular. Learning is really about the increased probability of a behaviour based on reinforcement which has taken place in the past, so that the antecedents of the new behaviour include the consequences of previous behaviour.

The theory of B.F. Skinner is based upon the idea that learning is a function of change in overt behavior. Changes in behavior are the result of an individual's response to events (stimuli) that occur in the environment. A response produces a consequence such as defining a word, hitting a ball, or solving a math problem. When a particular Stimulus-Response (S-R) pattern is reinforced (rewarded), the individual is conditioned to respond. The distinctive characteristic of operant conditioning relative to previous forms of behaviorism (e.g., Thorndike, Hull) is that the organism can emit responses instead of only eliciting response due to an external stimulus.

Reinforcement is the key element in Skinner's S-R theory. A reinforcer is anything that strengthens the desired response. It could be verbal praise, a good grade or a feeling of increased accomplishment or satisfaction. The theory also covers negative reinforcers -- any stimulus that results in the increased frequency of a response when it is withdrawn (different from adversive stimuli -- punishment -- which result in reduced responses). A great deal of attention was given to schedules of reinforcement (e.g. interval versus ratio) and their effects on establishing and maintaining behavior.

One of the distinctive aspects of Skinner's theory is that it attempted to provide behavioral explanations for a broad range of cognitive phenomena. For example, Skinner explained drive (motivation) in terms of deprivation and reinforcement schedules. Skinner (1957) tried to account for verbal learning and language within the operant conditioning paradigm, although this effort was strongly rejected by linguists and psycholinguists. Skinner (1971) deals with the issue of free will and social control.

Operant conditioning has been widely applied in clinical settings (i.e., behavior modification) as well as teaching (i.e., classroom management) and instructional development (e.g., programmed instruction). Parenthetically, it should be noted that Skinner rejected the idea of theories of learning (see Skinner, 1950).

EXAMPLE:

By way of example, consider the implications of reinforcement theory as applied to the development of programmed instruction (Markle, 1969; Skinner, 1968)

1. Practice should take the form of question (stimulus) - answer (response) frames which expose the student to the subject in gradual steps
2. Require that the learner make a response for every frame and receive immediate feedback
3. Try to arrange the difficulty of the questions so the response is always correct and hence a positive reinforcement
4. Ensure that good performance in the lesson is paired with secondary reinforcers such as verbal praise, prizes and good grades.

Principles:

1. Behavior that is positively reinforced will reoccur; intermittent reinforcement is particularly effective
2. Information should be presented in small amounts so that responses can be reinforced ("shaping")
3. Reinforcements will generalize across similar stimuli ("stimulus generalization") producing secondary conditioning

Cognitive behavioral therapy (or cognitive behavioral therapies or CBT) is a psychotherapeutic approach that aims to solve problems concerning dysfunctional emotions, behaviors and cognitions through a goal-oriented, systematic procedure. The title is used in diverse ways to designate behavior therapy, cognitive therapy, and to refer to therapy based upon a combination of basic behavioral and cognitive research.

There is empirical evidence that CBT is effective for the treatment of a variety of problems, including mood, anxiety, personality, eating, substance abuse, and psychotic disorders. Treatment is often manualized, with specific technique-driven brief, direct, and time-limited treatments for specific psychological disorders. CBT is used in individual therapy as well as group settings, and the techniques are often adapted for self-help applications. Some clinicians and researchers are more cognitive oriented (e.g. cognitive restructuring), while others are more behaviorally oriented (in vivo exposure therapy). Other interventions combine both (e.g. imaginal exposure therapy).

CBT was primarily developed through a merging of behavior therapy with cognitive therapy. While rooted in rather different theories, these two traditions found common ground in focusing on the "here and now", and on alleviating symptoms. Many CBT treatment programs for specific disorders have been evaluated for efficacy and effectiveness; the health-care trend of evidence-based treatment, where specific treatments for symptom-based diagnoses are recommended, has favored CBT over other approaches such as psychodynamic treatments. In the United Kingdom, the National Institute for Health and Clinical Excellence recommends CBT as the treatment of choice for a number of mental health difficulties, including post-traumatic stress disorder, OCD, bulimia nervosa and clinical depression, and for the neurological condition chronic fatigue syndrome/ myalgic encephalomyelitis.

The roots of CBT can be traced to the development of behavior therapy in the early 20th century, the development of cognitive therapy in the 1960s, and the subsequent merging of the two. Behavior therapeutical approaches appeared as early as 1924, with Mary Cover Jones' work on the unlearning of fears in children. In 1937 Abraham Low developed cognitive training techniques for patient aftercare following psychiatric hospitalization. Low designed his techniques for use in his organization, Recovery International, which supports people recovering from mental illness. Although Recovery

International was originally led by Low, he later adapted the techniques for use in lay-run self-help groups operating under the same name.

It was during the period 1950 to 1970 that CBT became widely utilized, with researchers in the United States, the United Kingdom and South Africa who were inspired by the behaviorist learning theory of Ivan Pavlov, John B. Watson and Clark L. Hull. In Britain, this work was mostly focused on the neurotic disorders through the work of Joseph Wolpe, who applied the findings of animal experiments to his method of systematic desensitization, the precursor to today's fear reduction techniques. British psychologist Hans Eysenck, inspired by the writings of Karl Popper, criticized psychoanalysis in arguing that "if you get rid of the symptoms, you get rid of the neurosis", and presented behavior therapy as a constructive alternative. In the United States, psychologists were applying the radical behaviorism of B. F. Skinner to clinical use. Much of this work was concentrated towards severe, chronic psychiatric disorders, such as psychotic behavior. and autism Although the early behavioral approaches were successful in many of the neurotic disorders, it had little success in treating depression. Behaviorism was also losing in popularity due to the so-called "cognitive revolution". The therapeutic approaches of Albert Ellis and Aaron T. Beck gained popularity among behavior therapists, despite the earlier behaviorist rejection of "mentalistic" concepts like thoughts and cognitions. Both these systems included behavioral elements and interventions and primarily concentrated on problems in the present. Albert Ellis's system, originated in the early 1950s, was first called rational therapy, and can arguably be called one of the first forms of cognitive behavioral therapy. It was partly founded as a reaction against popular psychotherapeutic theories at the time, mainly psychoanalysis. Aaron T. Beck, inspired by Albert Ellis, developed cognitive therapy in the 1960s. Cognitive therapy rapidly became a favorite intervention technique to study in psychotherapy research in academic settings. In initial studies, it was often contrasted with behavioral treatments to see which was most effective. During the 1980s and 1990s, cognitive and behavioral techniques were merged into cognitive behavioral therapy. Pivotal to this merging was the successful development of treatments for panic disorder by David M. Clark in the UK and David H. Barlow in the US.

Concurrently with the contributions of Albert Ellis and Beck, starting in the late 1950s and continuing through the 1970s, Arnold A. Lazarus developed what was arguably the first form of broad-spectrum cognitive behavioral therapy. He later broadened the focus of behavioral treatment to incorporate cognitive aspects. When it became clear that optimizing therapy's effectiveness and effecting durable treatment outcomes often required transcending more narrowly focused cognitive and behavioral methods , Arnold Lazarus expanded the scope of CBT to include physical sensations (as distinct from emotional states), visual images (as distinct from language-based thinking), interpersonal relationships, and biological factors.

Samuel Yochelson and Stanton Samenow pioneered the idea that cognitive behavioral approaches can be used successfully with a criminal population. They are the authors of, Criminal Personality Vol.I. This book has an extensive amount of information regarding the dynamics of criminal thinking and application of cognitive behavioral approaches.

CBT includes a variety of approaches and therapeutic systems; some of the most well known include cognitive therapy, rational emotive behavior therapy and multimodal therapy. Defining the scope of what constitutes a cognitive-behavioral therapy is a difficulty that has persisted throughout its development.

The particular therapeutic techniques vary within the different approaches of CBT according to the particular kind of problem issues, but commonly may include keeping a diary of significant events and associated feelings, thoughts and behaviors; questioning and testing cognitions, assumptions, evaluations and beliefs that might be unhelpful and unrealistic; gradually facing activities which may have been avoided; and trying out new ways of behaving and reacting. Relaxation, mindfulness and distraction techniques are also commonly included. Cognitive behavioral therapy is often also used in conjunction with mood stabilizing medications to treat conditions like bipolar disorder. Its application in treating schizophrenia along with medication and family therapy is recognized by the NICE guidelines (see below) within the British NHS.

Going through cognitive behavioral therapy generally is not an overnight process for clients. Even after clients have learned to recognize when and where their mental processes go awry, it can in some cases take considerable time or effort to replace a dysfunctional cognitive-affective-behavioral process or habit with a more reasonable and adaptive one.

3

Functional Significance of Behavior

Scientists have long sought to view the great diversity of organisms as a collection of distinct units. The species, as the atomic unit of diversity, represents a group of interbreeding natural populations that are reproductively isolated from other such groups. When individuals breed offspring, the genes of individuals are shuffled within a common gene pool representing the species' identity. The identity of species is based on the ability to breed, rather than on physical similarity. Limited transfer of genes between species causes different species to take on specific appearances and characteristcis.

SELECTION AND MICROEVOLUTION

Individuals will generally show great differences in the number of offspring they mange to contribute to the populations next generation. If differences in individual genotypes affect the bearer's reproductive success, then the frequencies of the genotypes will change over generations; genotypes with higher fitness become more common.

Fitness

Fitness describes an individual's ability to reproduce successfully, relative to that of other members of its population. In other words have a large number of feasible offspring. It measures changes in the proportion of genes from one generation to the next. Genes associated with higher fitness become more common while reduced fitness will decrease theit proportion. Fitness of a genotype is an averaged quantity which reflects the reproductive outcomes of all individuals with that genotype. When fitness measures the relative quantity of gene copies present in individuals of the next generation, they may have arrived there in one of several different ways. An individual may show greater reproductive success because it itself reproduced, or it may have helped relatives with similar genes to reproduce.

Artificial Selection and Breeding

Consider the work of a professional dog breeder. Lets consider a person trying to obtain a breed that is best suited for hunting of water fowl. The breeder specifically chooses individuals which most closely match the desired traits and selectively breeds them to each other in the hope of obtaining progeny with a combination of useful traits. Such plans can only succeed if:

- Phenotypic heritability in morphological and behavioral features is coded by <genes> (e.g. webbed toes, willingness to swim at least partially controlled by genes)
- Phenotypic variability is present (some have a particular characteristic, some don't)
- Differential survival/reproductive success of phenotypic variants is produced (you pick the ones who do)

Repeated, selective breeding events will alter the proportion of different genes over time. Genes which were present in those individuals that the breeder selected for reproduction will become overrepresented while those that occurred predominantly in discarded individuals will decrease in proportion. Animal breeding is a slow process, however, a combination of strong selection and a high degree of heritability can change the relative abundance of genes in a population up to 10% per generation.

Not all of the selections that we make are done intentionally. Penicillin became widely available during the second world war for the treatment of infected wounds. Just four years after drug companies began mass-producing penicillin in 1943, microbes began appearing that could resist it. With frequent and indiscriminant use of antibiotics we have fostered the emergence of antibiotic resistance in a variety of microorganism. The ability to withstand the effects of an antibiotic occurs when a rare mutation renders a small subset of individuals with lowered sensitivity to the effects of the drug. During the course of antibiotic treatment the wild type individuals are killed first as intended while the mutants are able to survive a bit longer. If the treatment is stopped before the drug had an opportunity to kill all pathogens regardless of slight differences in sensitivity to it, only resistant individuals will survive and be able to infect new hosts. With each antibiotic treatment that ends prematurely we unintentionally select for those individuals that exhibit a higher ability to tolerate the drug.

Natural Selection

Natural selection is the process by which favorable traits that are heritable become more common in successive generations of a population of reproducing organisms, while the proportion of less successful traits shrinks. "Favorable" and "successful" are defined in a purely functional sense as environmental conditions allow some individuals to leave more offspring than others. Selection cannot act on the genes per se but rather works on their expression into observable characteristics. Individuals with favorable phenotypes are more likely to survive and reproduce then the genotypes associated with favorable phenotyps will increase in frequency with each generation. Over time, this process can result in the emergence of adaptations that specialize a group of organisms for a particular ecological set of conditions (i.e., microevolution). If different subsets of the population adapt to excel in one of multiple, distinct niches, the emergence of any reproductive barriers between the groups may allow the population's split into separate species (i.e., macroevolution).

More offspring are produced than finite resources can support. Individuals thus can be viewed as in a constant struggle for existence. Individuals within a population are rarely clones, they commonly show variation in phenotypes as well as genotypes. Some of this variation in behavioral phenotypes is heritable. As successful variants are more likely to survive and reproduce, their genotypes will be become overrepresented in the next generation

Density-independent and density-dependent growth models: Exponential Model: a species can potentially increase in numbers according to a geometric series -- Thomas Robert Malthus (1766-1834) Logistic Model: the rate of population increase may be limited, i.e. it may depend on population density -- Pierre Verhulst (1838). Carrying Capacity (K): an environment's maximum persistently supportable load (Catton 1986).

Natural selection is the process by which environmental effects lead to varying degrees of reproductive success among individuals of a population of organisms with different hereditary characters, or traits. The characters that inhibit reproductive success decrease in frequency from generation to generation. It is the process whereby certain genes (alleles) gain greater representation in the following generations compared to other alleles. Adaptations are the complement of traits that increases the fitness of the owner. An individual's Fitness or Reproductive Success is the relative probability that an animal of a particular genotype and phenotype will manage to contribute its genes to the next generation

Calculations: Moths in Polluted (dark) Woodland

Dotted Whitish Form: (30/50) x 100 = 60% Melanic Form: (20/50) x 100 = 40%

60% - 40% = 20% The Dotted Whitish form is predated on 20% more than the melanic form in the polluted woodland.

Moths in Non-polluted (pale) Woodland

Dotted Whitish Form: (15/50) x 100 = 30% Melanic Form: approx. (30/50) x 100 = 60%

60% - 30% = 30% The Melanic form is predated on 30% more than the dotted whitish form in the nonpolluted woodland.

Their studies showed that the dotted whittish form was preyed on 20% more than the melanic form in the polluted woodland. The melanic form is predated on 30% more than the dotted whitish form in the nonpolluted woodland.

In addition to the difference in predation, their studies also showed that in both the polluted and nonpolluted environments the moths that were located on the limb joints instead of the trunks were less likely to be preyed on.

Howlett and Majerus showed that the dotted whitish form was at a higher risk of predation when resting against the polluted (dark) trees. The melanic form had a greater chance of survival and reproduction because they were less likely to be detected against the soot-covered trees, the same findings of Kettlewell years before. Their experiment proved that the melanic form was an anti-predator adaptation, which is why the dotted whitish form had become so rare. The melanic form had become the dominant trait for survival and reproduction in the polluted woodland.

The relationship between prey and predators continually changes. Prey need to find ways to outsmart predators in order to survive and reproduce. One way monarch butterflies increase their fitness is by forming huge groups of up to ten million. Increasing group size lowers the probability that any one monarch butterfly will be captured. This is known as the "dilution effect." Moreover, butterflies located in the center of a large group are more likely to survive than those on the outside. Monarch butterflies that choose to migrate to closed-area overwintering sites are less likely to be attacked by a predator. Also, by reacting as a group to the movement of a predator, monarch butterflies are better able to scare away predators. This "mass startle effect" is thought to stun the predators and provide time for the butterfly to escape (true shock and awe indeed). The aforementioned are collective ways in which the butterfly behaves in order to elude a predator. There are, however, certain individual inherent features that the monarch butterflies possess that increase their probability of avoiding a predator.

Monarch butterflies contain chemicals that are toxic to many predators. Evidently, this makes other similar but harmless species envious. In Batesian mimicry, a palatable species attempts to mimic an unpalatable species in an attempt to increase its own fitness. The monarch butterfly species is one that some Batesian mimics model themselves after perhaps because the monarch butterflies are so successful at avoiding predators. The Batesian mimics, although they are not harmful to predators, experience increased fitness because they model a potentially harmful species such as the monarch butterfly. Although monarch butterflies do not gain an increase in fitness as models of Batesian mimics, they do benefit from Müllerian mimicry. In Müllerian mimicry, two unrelated toxic species converge on a similar morphology. If more than one unpalatable species has a similar morphological trait, then the predators may more easily recognize a Müllerian mimic as potentially harmful. This saves both the mimics and the predators time and energy.

For many years the viceroy butterfly (Limenitis archippus) was thought to be a Batesian mimic of the Monarch. As it turns out, the viceroy can, depending on its diet, be more unpalatable than the monarch, thus both species benefit from resembling the other because a predator eating either one may become ill and avoid both in the future. Thus both species are examples of Müllerian mimicry.

Monarch butterflies display aposematism or warning coloration. This warning coloration is meant to be very conspicuous. Monarch butterflies make themselves conspicuous by having bright orange areas on its wings. Predators quickly learn that prey containing these bright colors are potentially harmful. For example, when a blue jay consumes a monarch butterfly, it vomits shortly after. From that point on, the blue jay associates features of the monarch butterfly, such as its bright colors, as unpalatable. Because of their morphological features, Batesian mimics, Müllerian mimics, and many other aposematic species all gain from the monarch's unpalatability.

Sexual Selection

Sexual selection, a subcategory of natural selection, was first recognized by Charles Darwin and occurs when individuals differ in their ability to compete with others for mates or to attract members of the opposite sex. By heavy courtship, fighting, or large territorial possession, males heavily compete for females. Even though a male may win a fierce competition for the mate of his choice, it is ultimately the female who decides on a partner that she wants. The female is often successful in her attempts to control reproduction by being choosy and having particular preferences for a male mate.

Females typically have a relatively larger investment in producing offspring. While a single male can often fertilize all of a female's eggs, she will not produce more offspring by mating with more than one male. By and large, a male's reproductive success increases with the number of mating opportunities he obtains, whereas a female's reproductive success is largely limited by how many offspring she can physically bear. This results in sexual selection, in which eager males compete with each other, and females become choosy in which males to mate with. Bateman's studies on mating behavior in fruit flies suggested that the origins of the unequal investment reside in the fact that male gametes (sperm) are cheaper to produce than eggs. Animals are therefore fundamentally polygynous, as a result of being anisogamous.

A female can have only a limited number of offspring, whereas a male can have a virtually unlimited number, provided that he can find females willing to mate with him. Thus females generally need to be much choosier about who they mate with.

A male can easily produce sperm in excess of what it would take to fertilize all the females that could conceivably be available [...] Hence the development of the masculine emphasis on courtship and territoriality or other forms of conflict with competing males.

In most animals the fertility of the female is limited by egg production which causes a severe strain on their nutrition. In mammals the corresponding limiting factors are uterine nutrition and milk production, which together may be termed the capacity for rearing young. In the male, however, fertility is seldom likely to be limited by sperm production but rather by the number of inseminations or the number of females available to him... In general, then, the fertility of an individual female will be much more limited than the fertility of a male... This would explain why in unisexual organisms there is nearly always a combination of an undiscriminating eagerness in the males and a discriminating passivity in the females.

Among polygynous species, the variance in male reproductive success is likely to be greater than the variance in female reproductive success.

The female, with the rarest exceptions, is less eager than the male... she is coy, and may often be seen endeavouring for a long time to escape.

Intersexual Choice: Why does she consider the trait "attractive"?

Females choose mates based on many factors. One important factor is male adornments, or ornaments. For example, Marion Petrie and Tim Halliday concluded that the removal of eyespots from a peacock's tail significantly reduces his attractiveness to females. Thus, females are extremely conscious of the visual stimuli provided by males. Other factors involved in female preference in mate selection include body coloration or "gifts" the male may present to the female before copulation. A mother is expected to always want the best for her child, even if it is a future child. Females select mates with certain traits, because they want their children to be healthy, viable, and reproductively successful. A direct and heritable benefit of mate choice by females is the assurance of bearing offspring that will survive well and display high general fitness. Non-genetic benefits of mate choice include fecundity advantages, food, parental care, or a good territory. All of the benefits, both heritable and nonheritable, ultimately lead to the greater survival of a female's offspring.

Passive Attraction Theory: Sensory Bias, conspicuous signals make an individual more likely to attract the attention of a mating partner. Elaborate sensory cues alert the female that the male is reproductively superior to others. Male adornments may more readily elicit the mating responses of some females who will thus mate preferentially in favour of the adorned males.

Nuptial gifts: spermatophore; territory, protection, resources; courtship indicative of parental investment. e.g., Dung beetle

The Good Genes Hypotheses regards assessment of the mate's state of health as a general indicator of viability and quality. Females exhibit mate choice in order to provide their offspring with a partner's genes that will advance their offspring's chances of survival or reproductive success. The healthy mate theory occurs when females prefer males healthy enough to produce and maintain elaborate ornaments. A good example of this is in female house finches, who choose male mates based on their bright coloration. Bright coloration tells the female that the male is more resistant to pathogens and parasites. In Hamilton and Zuk's "Revealing Signal Theory" bright ornaments reveal a genuinely healthy individual in good condition. In conditions where parasites ralter male showiness and parasite resistance is largely inherited, females ought to choose those with bright coloration.

Zahavi's Strategic Choice Handicap Theory: the presence of a costly trait is indicative of otherwise good underlying genes that allow an individual to prosper despite this handicap

Fisher's Run-away Selection: Females that mate with attractive males are compensated for reduced fecundity by bearing attractive offspring with higher than average mating success. This "Sexy Sons Hypothesis" works by alligning the presence of a particular morphological characteristic in males with a preference for it in females. Choosy females create a positive feedback loop favoring both males with these attributes and females that prefer them.

GENETIC COMPATIBILITY: MHC LOCUS GENES IN HUMAN MATE CHOICE

The Good Parent Theory suggests that choosy individuals ought to select partners on the basis of how well they will care for their offspring, e.g. as the female searches for a paternal male.

Inbreeding

Inbreeding refers to an elevated amount of breeding between close relatives. The resulting increase in homozygosity of the population exposes recessive, deleterious traits in homozygous form and may lead to inbreeding depression where inbred individuals exhibit reduced health and fitness and lower levels of fertility.

Outbreeding

The introduction of genetic material from unrelated individuals into a breeding line increases genetic diversity, thus reducing the incidence of disease or genetic abnormalities. It is used in breeding to restore vigor, size and fertility to a breeding line.

Sexual Selection and Ornamentation in Males

Sexual selection refers to the process by which changes in gene frequencies result from individuals that are better than others at either competing for or at attracting mates -- it is the evolution of traits based on differences in mating success among individuals if (1) some traits increase the ability to compete with individuals of the same sex for access to mates, or (2) some traits increase the ability to attract individuals of the opposite sex. Ornamentation in males, which are commonly the competing sex, may result from different evolutionary forces. Sexual selection is always harsher on the competing sex.

Species recognition Ornamentation in breeding individuals may serve in a quick recognition of others as members within a pool of genetically compatible breeding partners.

Intrasexual Competition and interference

Dominant individuals often gain preferred access to mates, desirable territory, or other advantages that will enhance the individual's chances for transmitting its genes into future generations. Precopulatory: A common form is based on an individual's ability to physically dominate a rival. In situations where groups of mates can be readily monopolized this will likely lead to increasing size in the competing sex. Fur seal females, which rely on a small stratch of beach for giving birth, can be monopolized easily in harems.

Winning in ritualized contests, producing a louder signal, or masking an opponent's call; dominance in social groups; territorial exclusion; alternative mating strategies; sexual interference.

Postcopulatory:

- Mate guarding;
- Anti-aphrodisiacs;
- Mating plugs; partners remain attached e.g. wolves;

Bruce effect is a form of pregnancy disruption when a female mammal is exposed to an unknown male. Examples may also include post-implantation failure in many species of rodents.

Infanticide:

- Sperm competition: e.g. primate or bat mating systems
- Basic Models of Molecular Evolution
- Adaptational Models

Adaptational Models are rooted in the view that natural selection plays the dominant role in structuring an organism's morphology, physiology, and behavior. The extant phenotype thus is the product with adaptations that have its possessor allowed to reproduce more successfully than individuals with alternative traits.

Neutralist Models

Alternative accounts for the evolution of traits focuses on genetic drift as a primary determinant for characteristics ranging from genes to complex traits. This acknowledges a significant role for historical contingencies, where present phenotypes of population members are largely determined by the genetic makeup of a small number of founding individuals. Although this view does not deny the role of natural selection in determining the course of adaptive evolution, the theory attributes a large role to genetic drift. When one compares the genomes of existing species, the vast majority of molecular differences do not influence the fitness of the individual organism. As a result, the theory regards differentiation in such features as neither subject to, nor explicable by, natural selection. Theoretical considerations suggest that such a process is most effective in small populations and when selecting among traits that are largely adaptively neutral.

Speciation and Macroevolution

Speciation is the evolutionary process by which new biological species may arise. Natural speciation has taken place over the course of evolution, although the relative importance of different mechanism in driving biodiversity remains a subject of debate. Debate also relates to the rates at which speciation events may occur over geologic time. Although some evolutionary biologists claim that speciation events have remained relatively constant over time (i.e., Phyletic Gradualism), palaeontologists such as Niles Eldredge and Stephen Jay Gould have argued that species may remain unchanged over long periods of time with speciation occurring in relatively brief intervals (i.e., Punctuated Equilibrium).

Allopatric Speciation

In allopatric speciation, two subgroups of a population become geographically isolated from each other (i.e., allopatric populations) through geographical or geological events. As these isolated populations undergo genotypic and phenotypic divergence, they are subjected to different selective pressures and may undergo genetic drift. New species form when the populations evolve independently such that they are no longer able of exchange genes when they come back into contact.

Peripatric and Parapatric Speciation

New species are formed when isolated, small peripheral populations split from the main population and occupy a novel ecological niche. Since small populations often undergo bottlenecks, genetic drift may drive the latter into significant genetic divergence (i.e., peripatric specieation). If geography only accounts for a partial separation, individuals of each species may come in contact across the barrier from time to time, but reduced fitness of heterozygotes leads to selection for reproductive barriers that prevent breeding between the two species.

Sympatric Speciation

In symptric speciation species diverge while inhabiting the same place. This may happen when a diploid cell undergoes failed meiosis, producing diploid gametes, and self-fertilizes to produce a tetraploid zygote. Alternatively, subgroups of a species may become dependent on different host organisms within the same area. With reduced fitness of heterozygotes, the emergence of reproductive barriers will drive the groups towards two distinct species.

Population Genetics and Behavior

Population genetics described the statistical distribution of genes in a particular breeding population, such as a breed of dog, and examines the role of various biological effects on gene distribution. While ordinary genetics predicts phenotypic makups of the next generation, population genetics concerns itself with genetics predictions for a breeding group several generations into the future.

Population

A population is the collection of organisms of a particular species within a given geographic area. When all individuals in a population are identical with regard to a particular phenotypic trait they are known as monomorphic. When the individuals show several variants of a particular trait they are polymorphic. The Gene pool for a species or population is the complete set of unique alleles present in it. While large gene pools indicate robust populations, extensive genetic diversity, and a good chance to survive bouts of intense selection, low genetic diversity is associated with reduced biological fitness and an increased chance of extinction.

Mutations

Mutations are alterations in genetic code ranging from changes to individual nucleotide bases along the DNA to largescale rearrangements of chromosomes. Mutations are rare and even at maximum rates allele frequencies will only change from p=1 to p=0.9993 per generation.

Genetic Drift

With small effective population size, random events can take on increasingly prominent roles. This focuses on changes in the genetics of the population as a result of chance alone.

Bottlenecks

Founder Effect

Examples include the relatively high rate of an autosomal recessive skeletal dysplasia (Ellis-van Creveld Syndrome) in the Amish population of Lancaster, Pa. Resulting in short-limbed disproportionate dwarfism, its incidence within the population is approximately 5 per 1000 live births and 2 per 1000 living persons. It can be traced to the initial immigrants Samuel King and his wife.

Selection

Selection: Strong selection for traits with high heritability can change proportions in excess of 10% per generation.

Non-random Mating

Non-random Mating: Mate Choice with assortative mating (e.g., blue and white form of snow geese), Positive assortative mating: mating among individuals who share particular genes or phenotypes. Negative assortative mating: preference for mating among individuals who are unlike each other with respect to particular genes or phenotypes. Inbreeding and Outbreeding. Inbreeding: refers to (preferential) mating between biological relatives. As relatives (i.e., ancestors of the first individuals are shared with those of the second individual), they carry genes which are "identical by descent"; extreme inbreeding: mating between sibs, half-sibs, parent-offspring. Individual inbreeding coefficient (i.e., Pedigree inbreeding): F represents the probability that the offspring is homozygous due to identity by descent (ibd) at a randomly chosen autosomal locus, ranges in value from 0 (no locus ibd) to 1 (all loci ibd). Significant factors for pedigfree inbreeding are

Outbreeding: (preferential) mating between non-relatives

Inbreeding depression: overrepresentation of homozygotes; e.g. matings between cousin are rare (0.05% in U.S.) but account for 20% of albinos; deficiency in immune systems; recessive, deleterious alleles occur more likely in homozygous condition; humans have on average 8 recessive, deleterious alleles. 1% increase in inbreeding produces 11% increase in the incidence of genetic diseases; Even small amounts of pedigree inbreeding in a random mating population greatly increase the incidence of some types of genetic disease. This is of particular concern in groups with small effective population sizes - Amish, Ashkenzi Jews.

MODELING ADAPTIVE BEHAVIOR

It is a daunting (or impossible) task to study behavior by simultaneously considering all significant factors and interactions that may impact it. An alternative method attempts to reduce the associated complexities by wrapping ones expectations into a set of mathematical abstractions based on a reduced number of significant variables. Resulting quantitative predictions are necessarily incomplete but are often less ambiguous and easier to test. The goal is to model an animals rational decisions of what action to take, given some information about the world. The player then faces consequences for the decision as a function of the action, the actions of others (if applicable), and the state of the surrounding environment. The player is expected to be rational by maximizing the expected utility.

Optimality Models

Optimality models quantitatively predict the consequences for a particular animal to behave in a certain way when pitted against the environment. It assumes that animals should behave so as to maximize their fitness. Optimality models consider separate, dependent, fitness benefits (B) and fitness costs (C) over a given range of values in a decision variable (i.e., the variable of interest). The solution to an optimality model attempts to find the point where the net benefits (i.e., B-C) are maximized. Individual solutions consider the consequences for a focal animal regardless of what strategies other individuals are relying on.

The main advantages of optimality models are in that they:

- Help us clarify our assumptions
- Force us to think about the information the animal has
- Tie many different ideas together in one place
- Generate testable predictions

Consequences of behavioral decisions are described as equations where the success of decisions of one individual do not depend on what decisions other animals make.

Decision Variable is the variable that we aim to optimize (e.g., optimize speed to obtain the best mileage when driving from here to there)

Currency refers to the criterion used to examine the outcome for different values of the decision variable (mileage)

Constraints are factors that limit the relationship between decision variable and currency (e.g., speed limits)

Optimality models use calculus and logic to solve for a minimum or a maximum in a specific function f(x). This can be achieved by taking the derivative of the function df[x)/dt (i.e., to obtain the slope of the function) and setting it to 0. Such an approach can also be used to explore sensitivity to changes in values of the decision variable and to determine what types of information are relevant for the relationship. Precise predictions can be formalized for empirical testing

Predator - Prey Population Cycles

In order to study predation and population oscillations, Huffaker used mite species, one being the predator and the other being the prey. By modifying the spatial structure of the habitat, he could manipulate the population dynamics and allow the overall survival rate for both species to increase. He did this by altering the distance between the prey and oranges (their food), establishing barriers to predator movement, and creating corridors for the prey to disperse. These changes increased habitat patches and in turn provided more areas for the prey to seek temporary protection. When the prey would go extinct locally at one habitat patch, they were able to reestablish by migrating to new patches before being attacked by predators. This habitat spatial structure of patches allowed for coexistence between the predator and prey species and promoted a stable population oscillation model.

NATURE AND NURTURE: FORMING ATTITUDES AND BEHAVIORS

How far is that human feelings and behaviors are inborn and how far are they all learned? This has been a very controversial issue over century and it is now come across as the nature versus nurture debate. "By birth the same, by custom different" is a quotation by Confucius which means that all human being are naturally alike, it is the habits that make each of every human being far apart and different from each other. Evolutional perspective which emphasizes on human kinship and cultural perspective which focuses on human diversity are the two main ideas that dominating the thinking of human similarities and differences in societies nowadays. Some scientists held the opinion that people behave as they do because of their genetic makeup. This is what we known as the "nature" of human behavior or attitude. While on the other hand, some scientist suppose that people think and act according to the way they are being taught and this is then known as the "nurture" of human behavior. A lot of research have been done and proved that both sides are partly right. Nature bequeaths us with innate traits and talent and nurture uses the environment factors to mold and shape these genetic tendencies. In my opinion, I think that human behaviors are largely shaped by the environment, in other word "nurture"' play a bigger part in designing our personality and behavior.

Evolutional psychology acknowledges that nature and nurture interact to form our behavior and personality. Our biological traits are not fixed blueprints, their expression rely on the environment factors. One of the classical examples is the experiment done by an American psychologist John Watson with a young orphan name Albert. The result shows that phobia could be explained by classical conditioning, an environmental learning. The little Albert started crying every time he see a furry object because he has learned that the furry object will always accompanying by a loud "Bang" sound that going to scare him. Other psychologist like B.F. Skinner, the father of behavioral science has also done experiments which produced dancing pigeons and proved that human behavior could be conditioned in similar way to the animal. Besides, a study published in "New Scientist" propose that sense of humor is not genetically inborn but a learned trait that induce by people and cultural. Obviously, environment definitely plays a role in shaping human's behavior and traits. If this is not the case, the identical twins should be exactly the same theoretically even they have been brought up in different place. But studies reveal that they are never exactly the same even though they are very much alike in most aspects.

Cultural diversity is one of the dominant factors that cause differences among people. There are millions of different culture exist across the world, cultural diversity even exist within nation. Malaysia is a very good example for cultural diversity within nation, despite the different of the physical look among Chinese, Indian and Malay, we are all living under the same environment and taking the same food and yet what make us so distinct from each other? Each of us practices different culture and tradition base on our race and religion, so base on different believe that we held, our behavior and thinking will never be the same even though we are all basically living under the same roof. On the other hand, the cultural diversity is even more obvious when people are from different part of the world. The language, customs, culture, and expressive behaviors that vary from one country to another country make people different from each other. And due to the factors above, people from different part of the world could hold different opinion about the same issue. Some societies do not wear shoes in house while other societies practicing it; some societies consider looking into people's eyes while talking is a show of politeness while some other societies consider such action as disrespectful. In addition to that, vary in parenting style in collectivist culture and individualistic culture provides a very clear example of cultural diversity between different societies. Parenting style is greatly influenced by the cultural context of a particular society because the cultural context is going to have impact on parents' moral values, believe, and socialization goals. So, what is consider normal in one culture maybe labeled as abnormal at another culture. Studies have shown that individualism and collectivism societies often hold different attitudes and value when teaching their child. The parenting style in collectivism societies emphasize on obedience, self-discipline and the importance to do well in school. Moreover, the children are expected to live with their parents until they get marry. While on the other side of the token, the parents in the individualism societies treat their children totally in an opposite way. They encourage their children to voice out their opinion and fight back whenever the children think they are correct. The parents in individualism societies, normally the westerners give much more freedom to their children compare to the parents in collectivism societies. Furthermore, the children of the westerners are expected to live with their own when they reach the age of 18 because they are considered as adult and must be able to stand on their own feet. Thus, we can imagine that how huge the different is it when people are being brought up with different culture.

Sometime, all the people behave in the same manner under certain situation; this has nothing to do with our biological trait but is due to the social norm. Social norms are also one of the factors that guide human behaviors and attitudes. Norms is the standards for accepted and expected behavior which society prescribe as "proper" behavior. Most of our behaviors are largely shape by the society and people seldom go against what the society expect us to do as we scared that we will be rejected by the society. For instance, stop when the traffic light turn red, open the door for an old lady, and keep quiet in the library. Try to imagine you speak loudly while everybody is silently studying in the library, what will happen? Definitely everybody will look at you one kind. Furthermore, there are also norms that we share across different culture. No matter what country are we in and what culture are we practicing, we respect our friend's privacy and do not divulge things said in confidence. This is known as universal friendship norms. Other than that, there are also universal status norms where people will talk in a more respectful way when talking to a higher-status people, for example lecturer and talk in a more casual way when the person is same level as you, for example friends. Lastly, the best known universal norms would be the taboo against incest. Sexual activity is prohibited among family member, neither parents with children, nor siblings with each other. Thus, people will govern their behavior regardless their genetic makeup in order to avoid rejection from the public.

Undoubtedly, genetic makeup makes a huge different between male and female. Male and female are born to be physically different, female is born to be has 70% more fat, 40% less muscle, 5 inches shorter and 40 pounds lighter than male. But biological traits only make different in sexes not gender. Gender is the characteristics people assign to male and female. Again, it is the society that assign specific gender role for men and women so that the society can function well with each of the gender carry out their assigned role and responsibility properly. Many people think that men are born to be more dominant and unemotional and women are born to be more emotional and soft which I do not think is true. Due to the gender role allocate to male and female, they are being brought up in a different way since they were young. Men are thought to be more independent, protective, and unemotional because since centuries ago men have been the head of the house, they are responsible to feed and protect the family; whereas women are thought to be more emotional, soft, and obedience because they are the one who going to give birth to baby, take care and teach of the child. They do not need to be strong because their job is stay and take care of everything in the house. Thus, it is the "nurture" that causes the two sexes behave in certain way and I think that men will stand out to be more dominant is due to the gender role that has been assign to them. Men are always given priority in the work place as their wages are always higher compared to the women who have same position as them and they are typical of those in higher-status positions. So, it is unassailable that men have higher status than women in the society and hence they are more dominant. On the other hand women are more emotional attached and express empathy more often is because they are being brought up and train in that way. Hence, the emotional trait inside women is not innate but learns through the families and environment.

In a nut shell, nature and nurture interact in forming our attitudes and behaviors. But just as I indicate the points above that in fact "nurture plays a bigger part than "nature" in determining human attitude and behaviors. If human behavior can be determine with merely biological traits, there will not have so many unsolved question and doubt about human thinking and behavior, nor the exits of psychology, sociology, anthropology to study the existence, culture, and behavior of human being. The environment and culture play a major role in the "nurture" part of molding our personality and behavior. If without the existence of cultural diversity, I believe that all of the people around the world will be much likely behave in the same manner as each and every one of us shares the same culture and the biological traits only play a minor role in determining our behavior traits and its major role is to differentiate us in term of our sexes. It is the same that apply to environment differences, as we are being raise up in distinct circumstance; therefore we act and behave differently. Because of the environment is changing over time, behavior and attitude of human are also constantly changing. Each and every one of us is unique due to environment and cultural that we are living in, so this is why that a same situation given but all of us react differently. In addition to that, people often choose and create their own situation base on their personal preference and believe. Thus, studying of human attitude and behavior is an ongoing process that will never end as long as human beings continue to survive in the earth.

Animal experimentation, also known as animal research and animal testing, is an experimental procedure in which animals are used within academic, scientific, research, biological and commercial establishments. It is when using live animals is considered legal, protected and regulated by law, because their ability to benefit humans is very important. Moreover, people of the world have improved their understanding about their lives, their health, the problems related to health and how to solve them, how to prevent them in the future by carrying out such experiments on animals. It could be argued that scientists over the world would not have developed e.g. their knowledge about animal behavior or probability of effects of products on humans ...show more content...

Animal experiments are carried out in many countries, such as the United Kingdom, the United States of America, Australia, Switzerland and Germany. However, experiments on almost 3 million animals are carried out in the UK every year, and this is the biggest number across Europe (Kolar, 2006). That is why, this essay will examine animal experimentation specifically in the UK and evaluate the suffering of animals in such experiments and provide possible solutions and ways to avoid or minimize suffering. In addition, the main questions that will be addressed in this essay are: 1) what experiments are carried out in the UK? and Why? 2) What are the requests for the researchers to perform such experiments in the UK? 3) How do animals suffer and in what ways? 4) How do laws and regulations, guidelines try to limit suffering? 5) How can suffering be further limited and minimized?

In the United Kingdom, animal experiments are carried out for different purposes: they may be scientific, medical, biological, biomedical and even commercial ones. Thus, in the UK all experiments on animals are divided into five major fields: 1) Basic (biological) research, 2) Disease research, 3) Drug development, 4) Regulatory testing of non-medical products (toxicity), 5) Educational purposes (Kolar, 2006). The following paragraphs will discuss more about them.

Firstly, basic research has helped scientists and researchers to increase the scientific... It may be high volume mass tourism or low volume and low impact tourism. It may generate high economic returns or low economic returns, be sustainable or unsustainable, domestic or international, and based on day visits or longer stays (Roe et al., 1997). Wildlife tourism includes activities such as bird watching, whale watching, reef diving, gorilla tourism and photographic safari. Wildlife tourism can include interaction with animals in captivity but for the purposes of this paper the narrower definition of wildlife tourism as denoting only human contact with animals in the wild is used.

Wildlife tourism is a holiday with a difference. A wildlife tourism holiday offers a different experience for each traveller on each occasion, as the major component of the holiday, the wildlife, is unpredictable. Tourists often feel that they, and they alone, have experienced a particular aspect of an animals behaviour.

Wildlife tourism often involves not only seeing rare or exciting animals in their natural habitat, but visiting areas that are difficult to travel to and hence not visited by many tourists. This perceived exclusivity adds to the appeal of wildlife tourism.

In the a study reported in the Times, three of the top five Things to Do Before You Die involved an aspect of wildlife tourism (swim with dolphins, whale watching and swim with sharks respectively). For many participants a wildlife holiday will be remembered for many years to come as a holiday with a difference.

Trends in Consumer Behaviour that are Catered for by Wildlife Tourism

Public interest in conservation and environmental matters is growing. 85% of industrialised world citizens believe that the environment is the most important public issue (Carson and Moulden, 1991). This socio-cultural trend has led to an increase in the demand for ecotourism. Ecotourism is nature-based tourism that involves ecologically sustainable management of natural areas and an interaction with or observance of either flora or fauna. This is precisely what wildlife tourism provides.

There has also been a trend towards experiential tourism; that is active, rather than passive travel. People increasingly want to feel that they are participating in and influencing the area they travel to. Wildlife tourism caters for this trend in that tourists often feed or otherwise interact with the wildlife they visit.

The major world tourism markets, Europe and North America, are affected by an increase in the proportion of older people in the population. In addition, changes in working practises have meant that people have more flexibility to get longer periods of time off work. The days when gap years were restricted to students are gone. More mature people than before are taking sabbaticals from work or travelling after retirement. Wildlife tourism caters well for these grown up gappers as a longer time frame will allow remote geographical areas to be accessed and increase the chances of a successful wildlife encounter.

Management Issues Facing Wildlife Tourism

The management issues facing wildlife tourism are twofold, identifying methods of minimising the impact on the wildlife resource itself and ensuring that the tourists receive a valuable experience from the holiday. Sometimes these two aims are mutually exclusive. There are many direct impacts of wildlife tourism on wildlife. They can be divided into categories such as habitat destruction, disturbance of feeding and disease transmission. Each of these categories will now be illustrated with an example.

Tourism can contribute to the destruction of wildlife habitat. In the Norfolk Broads, UK, large volumes of boat traffic, especially motor boats, result in considerable boat wash and river bank erosion and undermine on-going efforts to improve the Broads severely impaired water quality and threatened aquatic wildlife (Brouwer et al, 2001). The consequences of erosion can lead to important habitats being lost and the silt from the disappearing banks clogging up the waterways, which further deteriorates water quality.

In the Yacatan Peninsula, Mexico, 'boatloads of tourists were driven into groups of feeding flamingos to make them take flight' (Long, 1991, cited in Chin, Moore, Wallington & Dowling, 2000). This type of disturbance during feeding can have several effects depending on the magnitude of the disruption. Some birds may take flight temporarily, but return after the disturbance ends. Other birds may modify their feeding habits and desert the site of tourism disturbance permanently. When a bird is unexpectedly forced to take flight during feeding, energy intake terminates and energy expenditure significantly increases. Where disturbance causes a bird to desert a particular site, the availability of suitable alternative sites is critical for the bird's survival. Furthermore, the site in which it settles may already be populated or may be of lower quality resulting in lower rates of energy intake.

The possible transmission of diseases from humans to animals is a serious threat resulting from the close contact between species that many wildlife tourism experiences involve. It is though that the mountain gorillas of Rwanda are particularly affected. Humans and gorillas share about 97% of their genetic makeup and they are therefore susceptible to many of the same diseases. These include tuberculosis, influenza, measles, polio and intestinal parasites (Cameron). The single population of 300 mountain gorillas in the Virunga Volcanoes is particularly badly affected. More than seventy tourists and a similar number of guides, porters, rangers and researchers visit seventy percent of the gorillas in this population daily. There have been several outbreaks of disease that can probably be attributed to humans. This includes an epidemic in 1988 in which six habituated gorillas died of respiratory illness and twenty-seven more became ill.

Methods of managing these and other negative effects on wildlife caused by wildlife tourism include restricting the number of tours allowed, education and finally licensing, and thereby controlling, the activities of tour groups.

The most popular example of restricting the number of tours allowed is that of the Galapagos Islands in Ecuador, which were protected in 1934. In the Galapagos Islands parts of the islands are designated as 'Intensive Visitor Zones', where a maximum of ninety people are allowed simultaneously on shore. The 'Extensive Visitor Zones' are open to groups of less than twelve individuals (de Groot, 1983).

Education of operators and tourists about appropriate behaviour to adopt in the presence of wildlife is another method of managing the negative impacts of wildlife tourism on the animals involved. A code of conduct can be distributed to tour operators and publicly displayed in the local area.

Tourist education can develop a demand-lead requirement for responsible tour groups.

Licensing of tour companies is a method of minimising the impact of wildlife tourism. Licensing is often dependent on compliance with particular regulations, such as minimum approach distances. Many whale watching regulations specify a minimum approach distance of 100m (Brouwer et al, 2001).

The choice of type of management depends on the nature of the target species and the severity of the impacts on them. An analysis of the impacts of the specific wildlife tourism would be required before a management method was selected.

As well as managing the needs of the wildlife by minimising wildlife tourism's impact on them, tourism managers, of course, need to consider the needs of the tourists. Tourists want a wildlife watching experience that is both enjoyable and informative. The skills and knowledge of tour operators add to the experience of wildlife tourists and so should be maximised.

Wildlife tourism is inherently seasonal. Most species act differently according to the season. The best example of this is migration. The 'Great Migration' is the movement of over a million wildebeest between the Masai Mara and the Serengeti plains. There are numerous wildlife tourism opportunities presented by this, but there are of course dependent on the timing of the migration. The migration usually occurs May to June, but this can be affected by rainfall, which in turn affect availability of vegetation. Wildlife tourism management involves understanding the environmental influences on species behaviour and how this will affect tourism opportunities.

Wildlife tourists also want to get 'close up' to the wildlife (Orams, 2000). Managers of wildlife tourism need to attempt to ensure predictable occurrence of species within a relatively restricted area. Managers of wildlife tourism need to carefully integrate visitors and wildlife through management of their interactions whilst still ensuring damage to the species is limited.

Managers of wildlife tourism also need to ensure that there are other activities to keep tourists amused for times when wildlife does not co-operate with tour schedules. There needs to be an awareness of either other geographical locations that can be utilised to view the same wildlife or other species that can be more easily located.

HUMAN AND ANIMAL COMMUNICATION

Language is commonly described as a communication tool which allows humans to express their feelings, thoughts and helps them to understand the world. It defines the cultural background of a particular society and separates human beings from other animals. Therefore, human language is believed to be a distinguishing factor, which makes our species unique (Anderson, 2004). It is important to notice that all the other animals also communicate by using different structures of communication systems that are innate (Hirshon, date unknown). Animal communication lacks creativity, flexibility and is not as complex as human communication. This essay will discuss the main features that distinguish human language from the ways other animals communicate, its origins and finally it will look at different attempts to teach animals to acquire human language. The essay concludes that no animal is able to create complex sentences in a language, owing to its different anatomy of vocal tracts and inability to understand the meaning of words.

The scientists report that there are over 6,900 distinct languages in the modern world. Nevertheless, David Lightfoot, a linguist at Georgetown University in Washington, points out that because of human ability to speak and our anatomy of the vocal apparatus, only one language should be considered: human language (Hirshon, date unknown). It is in organized and complex form, structured by different sets of grammatical rules. Language is as important as breathing. It helps human beings to socialise, learn about past cultures and progenitors, develop themselves and pass the knowledge through the next generations. In fact, it is crucially important for speakers to have a knowledge of the grammar of a spoken language.

As far as the nature of language is concerned, many philosophers and scientists have debated this question for a long time, presenting contrasting ideas and concepts. Although a number of different theories have been displayed so far, there is still no direct evidence which tells us how the language has originated. It is certain that divine theory is one of the oldest thesis, suggesting the language as the gift of the God to mankind. Judeo Christians believe that Adam was given the power to name all the things that the world consists of. God communicated with the first people in a language that was easily understood by their brains (Yule, 2010). In terms of divine creation, many experiments have been carried out since the Ancient Times. For instance, an Egyptian pharaoh, Psammetichus, decided to place two infants in the mountains, isolating them from society. They were only in close contact to mute servants. Surprisingly, after several years they were able to say a word meaning 'bread' in a language which had not existed for a long time. Another interesting examination was carried out by James IV of Scotland thousands years after Psammetichus. Indeed, he repeated the same experiment, getting significant results: isolated children were capable of speaking Hebrew. Therefore, James IV formulated a thesis that Hebrew was the language Adam and Eve used in order to communicate in Eden (Fromkin, Rodman and Hyans, 2003).

On the contrary to divine source, another important thesis of this issue is evolutionary development. Many scientists suggest that the language has originated by the evolution of mankind. It resulted in the development of the speech production and the perception apparatus (Fromkin, Rodman and Hyans, 2003). Stephen Jay Gould (1994), as cited by Fromkin, Rodman and Hyans (2003, p.60), said that the language progressed 'step by step and is such an integrated "all or none" structure'. It is very probable that the brain expanded its size, which enabled the processing of different kinds of things. Thus, the evolution of human language must have been a long - lasting movement. The evolutionary thesis also claims that the development of human language was linked to the alterations in the brain and nervous system. However, when it comes to vocal tracts, there is no doubt that some species, such as talking parrots and mynah birds, have also experienced the development of vocal apparatus. As a result, these animals are able to distinguish different sound patterns, but they cannot acquire human language. It may be the case that vocal tracts are not so necessary in terms of human language development (Fromkin, Rodman and Hyans, 2003).

Another aspect of scientific research is the role of the brain lateralization (a concept, suggesting that the two halves of the brain`s cortex are responsible for different functions), which is closely related to the tool-making source. Yule (2010) indicates that making tools by early humans had a great influence on the development of the language since functions, controlling both speaking and object manipulation, are located in the left hemisphere of the brain.

An additional theory is the natural sound source. Its main idea is that natural sounds may have been imitated by unsophisticated words, which human ancestors heard around them. Certainly there are numbers of onomatopoeic words in every language, but considering, how most of the soundless things and abstract concepts in the world could not refer to natural sounds, make the scholars be sceptical about that theory (Yule, 2010).

The last possible thesis, discussed in the essay, will be the genetic source. In fact, many scientists believe that human offspring are born with an innate and unique capacity for language. In other words, any normal child, independently on its race, place of birth or economic heritage, is able to acquire a language to which he or she is exposed (Fromkin, Rodman and Hyans, 2003).

Regardless how much effort we make to discover the nature of human language, there is still no direct evidence that would prove any of the theories mentioned above. However, as Chomsky (1972, p.67, 68) says, 'human language appears to be a unique phenomenon, without significant analogue in the animal world'. Thus, what are the special features of the language which set humans apart from the other animals? The first uniqueness of human language is infinity capacity. Lightfoot claims that it enables humans to produce short sentences as well as long and complex ones. What is more, once the human learns a language, he is able to create new phrases and sentences. It is possible, only if he knows the rules of the grammar. In contrast, animals are not capable of this. Birds, for instance, can only sing on the basis of 'genetic programming', or what they have heard before (Hirshon, date unknown).

Furthermore, human beings are stimulus - independent, which means that they can talk about different situations without being determined to do so. This feature is closely related to displacement. Yule (2010) points out that human beings can talk about past, future and some imaginary situations. Animal communication is just a response in the immediate environment, connected with stimulus feeling of pain, the sight of predator or a desire of mate. A small exception for this are honeybees. In fact, they are able to inform the other bees about the location of nectar by performing a dance. However, it must be the most recent food source and they cannot relate to vertical distance.

Another difference between human and animal communication is creativity that enables human beings to produce new words, phrases and expressions. It is true that animals are born with a limited number of communication signals and cannot expand them in any way. As well as this, human language is culturally transmitted, acquired by close contact with a particular society and passed through generations, whereas animal communication represents biological transmission. It means that they are 'genetically programmed' to instinctively use a limited set of specific signals (Yule, 2010).

An important factor that separates human beings from other animals is also arbitrariness. Human language has signifier and signified. The meaning of particular words may change through the distance of time, but the point is that a word is a relationship between an image and its meaning (Anderson, 2004).

Human beings communicate with each other, using language as the main tool. However, the language should not only be considered as speech. Indeed, it also includes body gestures, verbal communication, facial expressions and sometimes specialised signs, especially when it comes to sign language that enables deaf people to communicate. It is a proven fact that these forms of communication are also common for animals. Although animals do not use sophisticated expressions and phrases to show their feelings and simply cannot talk in a way that humans do, they use different structures of communication systems that are innate. These systems usually involve distinct sounds, ranging from roars to whines, gestures and signals, such as alarm calls. It is said that communication allows animals to find food, avoid predators, mate as well as carry for their young (Castaldo, 1992).

Throughout the years, many scholars have tried to teach particular animal species to use human language as a tool of communication between humans and animals. Most of the projects involved apes that were taught by using different methods. Viki was one of the first apes that took part in the process of teaching the human language. The examination was carried out in the 1940s by scientist couple, Catherine and Keith Hayes, who raised Viki as if she had been their own child. The aim of this experiment was to make an ape say English words by shaping her mouth, while she was creating sounds. As a result, she could utter some poorly articulated words. In fact, the project brought a clear evidence that non-human primates are able to produce complex human sounds since they have different anatomy of vocal apparatus (Yule, 2010). The next investigation, aimed at discovering if an ape is able to learn one or more grammar rules, was carried out by Herbert Terrace in the 1970s. The psychologist involved an ape, called Nim Chimpsky, which was taught a version of American Sign Language. As Holmberg (date unknown) highlights, Terrace tried to compare Nim`s ability of communication with a two-year-old human child. He reached a conclusion that Nim acquired the basic linguistic knowledge, being able to use signs with their meanings and create new messages with regard to the rules he had learnt. However, Nim`s utterances were significantly different from that produced by human children. In fact, the ape constantly reversed the word order rules without any alteration in the meaning, whereas human children tend to follow the word order rules of adult language. Nim`s utterances were never longer than two words in terms of its meaning. Additional words, which seemed to be in an unplanned order, were always added to emphasise. Thus, Nim was able to learn many new words, but did not acquire grammar. He always created sentences that were related to requests for items of favours from humans (Holmberg, date unknown). Not only did the scholars teach apes to articulate human sounds correctly and use the sign language, but also to communicate using plastic shapes. Such an experiment was conducted by Ann and David Premack, who taught a chimpanzee, called Sarah, to use a set of plastic shapes and relate them to objects and actions. In the final stage she was able to request for some items, showing the understanding of the meaning of particular words (Yule, 2010).

Furthermore, when it comes to imitating human speech, scientists have proved that some parrots are able to produce noises, which may be heard as human sentences, but they cannot refer any word to its meaning. Anderson (2004) presents parrot Alex as an evidence for this thesis. Alex was raised by Irene Pepperberg and through the distance of time, he has learnt a vocabulary to name objects, shapes and numbers. Indeed, the communication system Alex has acquired could be characterised as 'an inventory of individually meaningful words rather than a set of holistically interpreted utterances' (Anderson, 2004, p.302). It cannot be said that the parrot has possessed human language as it could not understand the meaning of words it has uttered.

In terms of teaching animals to acquire human language, the research at the Dolphin Institute in Hawaii, involving bottlenose dolphins, should also be considered. Despite the fact dolphins lack vocal cords and cannot vocally imitate human language, the research has shown that they can understand the language semantically and syntactically (Herman, 2009).

This paper has discussed the reasons why the human language is so unique that it separates human beings from other animals. However, the language is just a tool used by humans to communicate and the communication is universal among all the species in the world. It should be noted that any of the scientific research, which has already been conducted, has not answered the question on the nature of human language, but it has been proved that no animal is able to possess human language. On the other hand, animal communicative skills may help humans to understand their behaviour. In the essay I focused on reviewing a number of different examinations on animals, aimed at teaching them to use human language. The question is if humans could possess the language of a particular animal species and would they be able to communicate with the animals using that language.

Nature vs. nurture has been discussed by philosophers in the past and by scientists most recently. Philosophers such as Plato argued that all knowledge was inherited through your parent and when you were told something you didn't learn it you were just reminded of it. Aristotle however argued that all humans were born with a blank slate and built on it with influence from there environment. In the 1700's the empiricists and the internalists took over the argument. They fought through letters explaining there point of views and denouncing the others. This leads to Pavlov coming up with the idea of behaviorism in the early 1900's. Behaviorism became the new wave of Psychology and influenced a lean towards the nurture side. It was not effectively argued against until 1928 when Watson published his book. This opened up the floodgates for environmental influences studies. Soon the idea of nurture was the popular excuse for behavior. Studies using animals were the most popular was in which scientists used to prove a theory, or disprove a theory. The newest studies use human twins to prove nature vs. nurture.

An age-old question has been asked for generations before us. What is the reasons behind the development of human behavior? There have been many theories formulated to explain why humans behave the way they do. Explanations vary from demonology to magnetic fluids controlling people's behaviors. Over time, two theories have remained popular in academic fields such as philosophy and psychology. The surviving theories for behavior stem from physiological and sociological explanations. However, the two explanations have not always been compatible with each other. The famous nature vs. nurture debate over human behavior resulted from conflicting views between proponents of the physiological (nature) and sociological (nurture) explanations. Throughout history, research has swayed popularity back and forth between the theories. Yet, theorists have broken down the line separating nature and nurture. Today, people us both explanations in research to advance the knowledge of human behavior.

- Thousands of years before the field of psychology, philosophers pondered on human behavior. As early as 350 BC, such philosophers as Plato and Aristotle tried to understand behavior. The question of nature or nurture as the primary drive can be traced to these times. Plato believed behavior and knowledge was due to innate factors. Author Fiona Cowie states, "The claim that the character of our mental furniture is to a large extent internally rather than environmentally determined found its first substantive defense in the works of Plato..." (Cowie, 1999, p. 3). Plato theorized, and Descartes later agreed, that all knowledge is present at birth. Plato also believed that the environment played a part in human processes, but he thought it had an unique role. He believed the environment did not teach people anything new, but its purpose was to remind people of information they already knew (Cowie, 1999). Although Plato's views are not supported today, he laid the groundwork for other researchers to follow.
- On the other hand, philosopher Aristotle theorized a different idea about human behavior. He presented the idea that humans are born into the world with a "blank slate" and people's behavior and thoughts are due to experience (Ashcraft, 1998). His tabula rasa explanation believed that the environment and experience were the important influences in human behavior. Unlike Plato, Aristotle hypothesized that humans were not born with knowledge, but they acquire it through experience (Ashcraft, 1998). Aristotle's idea of the tabula rasa is not believed today. Nevertheless, his belief that the environment was a vital factor in behavior influenced many empiricists throughout history.
- During the late 1700s, the nature vs. nurture debate began to heat up between philosophers. Internalists (nature) and empiricists (nurture) wrote literature back and forth trying to prove their beliefs and disprove the other's theories. Two philosophers, G.W. Leibniz and John Locke, were main representatives of their respected explanations. Leibniz promoted the internalism point of view.
- Cowie states, "...Leibniz's position on this issue is, of course, that the tabula is far from rasa: 'The soul inherently contains the sources of various notions and doctrines, which external objects merely rouse up...' " (Cowie, 1999, p. 7). Leibniz argued against Locke and other empiricists stated that "...there is no way ideas which come into the mind from outside can be formed into beliefs and judgments without the operation of specific internal mechanisms"

(Cowie, 1999, p. 17). At the same time, John Locke and his fellow philosophers campaigned for empiricism. Like Aristotle, the philosophers believed that humans' thoughts and actions were determined not by innate factors, but by the their unique experiences (Ashcraft, 1998). Locke argued against the internalists by examining different human processes such as logic and reasoning. He would ask how it was possible to use logic and reasoning if people were born with all of the knowledge they would ever acquire (Cowie 1999, p. 19). The contrasting views of the two groups had begun the nature vs. nurture debate, which would linger in the fields of philosophy and psychology for decades.

- A point should be made that even though the interalists and empiricists felt strongly about their theories, the explanations were not entirely opposite of each other. Cowie explains, "...rhetoric aside, both empiricists and nativists are both internalist and externalists about the origin of what is in our minds"(Cowie, 1999, p. 17). Even Leibniz and Locke stated that the philosophies sometimes were only different by the choices of words they used to describe their theories. Leibniz once wrote that fundamentally their views were the same about the nature vs. nurture question (Cowie, 1999).
- Over the next couple hundred years, popularity was split between nature and nurture. However, in the early 1900s Ivan Pavlov accidentally discovered what eventually became labeled as behaviorism. Behaviorists believed that the environment was the greatest factor in shaping behavior. The theory quickly gained notoriety in psychology and swayed popularity to the nurture side. One of the leaders in behavioral research was John Watson, who is most recognized for his work in conditioning "Little Albert." In 1928, Watson published a book that included his idea that infants were like clay. Watson stated that he could make an infant anything he desired by manipulating the environment (Barnet, 1998).
- Watson wrote, "Give me a dozen healthy infants...and my own special world to bring them up in and I'll guarantee to take any one...and train him to become any type of specialist..."(Amsel, 1989, p. 24). In the 1960s, Skinner also became well known for his research in behaviorism.
- Most of his work dealt with behavior modification with animals (Amsel, 1989). Probably Skinner's most famous research is when he conditioned pigeons to guide missiles (Modgil, 1987). However, Skinner, and others in his field, began to receive scrutiny for ignoring the biology of humans. Yet, Skinner responded by saying, "The objection to inner states is not that they do not exist, but that they are not relevant in a functional analysis" (Modgil, 1987, p. 228).
- From the 1920s to 1950s, behaviorism and nurturism dominated psychology. The domination did not go without any challenges, and it did not last forever. As early as 1929, behaviorism came under attack by psychologists who believed genes were the key to human behavior. To begin, Arnold Gesell questioned the environmental view of Watson. Author Myrtle McGraw states, "Gesell contended that there was nothing one could do through training young infants to accelerate their development; one simply had

to wait until the cells of the nervous system 'ripened'"(McGraw, 1995, p. 264). The biological research continued to build against behaviorism, and its popularity began to decrease.

- In 1959, the final attack that swayed popularity to the nature side of development originated with Noam Chomsky and other psycholinguists. Chomsky attacked behaviorism's scientific empiricism, especially dealing with the acquisition of language (Amsel, 1989). Ashcraft (1998) explains, "...Chomsky argued not only that the behaviorist account of language was seriously wrong and misguided, but that behaviorism was unable in principle to provide useful scientific knowledge of language" (p. 22). In addition, research developments in physiology and new studies involving genetics, such as adoption studies, and studies on twins, popularized genetic influence over environmental.

The most resent studies that have been done on twins and adoption use both identical and faternel twins. This consists in the studying of twins that were separated at birth and grew up in separate homes. Identical twins are 100% genetically similar and offer exact genetic replicas to study, where fraternal twins are the same as any other siblings at 50% similar (Vanderbilt pg6). Some of the final results of these studies show astonishing similarities between identical twins, yet others show little evidence of these similarities. With fraternal twins there is some similarities but none that are complete evidence of the nature theory. These studies fuel the pot for both the nature and the nurture ideas.

The nature vs. nurture debate over the last forty years has reached an agreement that they both influence the development of human behavior. In the 1960s, researchers from both theories began to study the interaction of the genes and the environment (Devlin 1997). Dr. Ann Barnet explains, "Even in an unborn baby, genes and environment interact almost from the moment of conception"(Barnet, 1998, p.10). The interaction between nature and nurture can be summed up by the statements of Dr. Fausto-Sterling and Dr. Evan Balaban. Fausto-Sterling states, "People want simple explanations for hard-core problems. If there was an antitestosterone drug that we could to inject to make young boys nice...it would be easier and cheaper than transforming schools...or whatever is at the heart of the problem" (Barnet, 1998). However, Balaban replies, "...don't hold your breath if you think looking for genes to help you understand violence. I would put my money on some clever environmental manipulations,

4

Habitat

A habitat is an ecological or environmental area that is inhabited by a particularspecies of animal, plant, or other type of organism. It is the natural environmentin which an organism lives, or the physical environment that surrounds a speciespopulation. A habitat is made up of physical factors such as soil, moisture, range oftemperature, and availability of light as well as biotic factors such as the availability of food and the presence of predators. A habitat is not necessarily a geographic area—for a parasitic organism it is the body of its host, part of the host's body such as the digestive tract, or a cell within the host's body.

MICROHABITAT

A microhabitat is the small-scale physical requirements of a particular organism or population.

Monotypic Habitat

The monotypic habitat occurs in botanical and zoological contexts, and is a component of conservation biology. Inrestoration ecology of native plant communities or habitats, some invasive species create monotypic stands that replace and/or prevent other species, especially indigenous ones, from growing there. A dominant colonization can occur from retardant chemicals exuded, nutrient monopolization, or from lack of natural controls such as herbivores or climate, that keep them in balance with their native habitats. The yellow starthistle, *Centaurea solstitialis*, is a botanical monotypic-habitat example of this, currently dominating over 15,000,000 acres (61,000 km^2) in California alone. The non-native freshwater zebra mussel, *Dreissena polymorpha*, that colonizes areas of the Great Lakes and the Mississippi Riverwatershed, without its home-range predator control, is a zoological monotypic-habitat example. Even though its name may seem to imply simplicity as compared with polytypic habitats, the monotypic habitat can be complex.

Habitat Destruction

Habitat destruction is the process in which natural habitat is rendered functionally unable to support the species present. In this process, the organisms that previously used the site are displaced or destroyed, reducing biodiversity. Habitat destruction by human activity is mainly for the purpose of harvesting natural resources for industry production and urbanization. Clearing habitats for agriculture is the principal cause of habitat destruction. Other important causes of habitat destruction include mining, logging, trawling and urban sprawl. Habitat destruction is currently ranked as the primary cause of species extinction worldwide. It is a process of natural environmental change that may be caused by habitat fragmentation, geological processes, climate change or by human activities such as the introduction of invasive species, ecosystem nutrient depletion, and other human activities mentioned below.

The terms habitat loss and habitat reduction are also used in a wider sense, including loss of habitat from other factors, such as water and noise pollution.

Impacts on Organisms

In the simplest term, when a habitat is destroyed, the plants, animals, and other organisms that occupied the habitat have a reduced carrying capacity so that populations decline and extinction becomes more likely. Perhaps the greatest threat to organisms and biodiversity is the process of habitat loss. Temple (1986) found that 82% of endangered bird species were significantly threatened by habitat loss. Endemic organisms with limited ranges are most affected by habitat destruction, mainly because these organisms are not found anywhere else within the world and thus, have less chance of recovering. Many endemic organisms have very specific requirements for their survival that can only be found within a certain ecosystem, resulting in their extinction. Extinction may take place very long after the destruction of habitat, however, a phenomenon known as extinction debt. Habitat destruction can also decrease the range of certain organism populations. This can result in the reduction of genetic diversity and perhaps the production of infertile youths, as these organisms would have a higher possibility of mating with related organisms within their population, or different species. One of the most famous examples is the impact upon China's Giant Panda, once found across the nation. Now it's only found in fragmented and isolated regions in the south-west of the country, as a result of widespread deforestation in the 20th Century.

Geography

Biodiversity hotspots are chiefly tropical regions that feature high concentrations ofendemic species and, when all hotspots are combined, may contain over half of the world's terrestrial species. These hotspots are suffering from habitat loss and destruction. Most of the natural habitat on islands and in areas of high human population density has already been destroyed (WRI, 2003). Islands suffering extreme habitat destruction include New Zealand, Madagascar, the Philippines, andJapan. South and east Asia — especially China, India, Malaysia, Indonesia, and Japan — and many areas in West Africa have extremely dense human populations that allow little room for natural habitat. Marine areas close to highly populated coastal cities also face degradation of their coral reefs or other marine habitat. These areas include the eastern coasts of Asia and Africa, northern coasts of South America, and the Caribbean Sea and its associated islands.

Regions of unsustainable agriculture or unstable governments, which may go hand-in-hand, typically experience high rates of habitat destruction. Central America, Sub-Saharan Africa, and the Amazonian tropical rainforest areas of South Americaare the main regions with unsustainable agricultural practices or government mismanagement.

Areas of high agricultural output tend to have the highest extent of habitat destruction. In the U.S., less than 25% of native vegetation remains in many parts of the East and Midwest. Only 15% of land area remains unmodified by human activities in all of Europe.

Ecosystems

Tropical rainforests have received most of the attention concerning the destruction of habitat. From the approximately 16 million square kilometers of tropical rainforest habitat that originally existed worldwide, less than 9 million square kilometers remain today. The current rate of deforestation is 160,000 square kilometers per year, which equates to a loss of approximately 1% of original forest habitat each year.

Other forest ecosystems have suffered as much or more destruction as tropical rainforests. Farming and logging have severely disturbed at least 94% of temperate broadleaf forests; many old growth forest stands have lost more than 98% of their previous area because of human activities. Tropical deciduous dry forests are easier to clear and burn and are more suitable for agriculture and cattle ranchingthan tropical rainforests; consequently, less than 0.1% of dry forests in Central America's Pacific Coast and less than 8% inMadagascar remain from their original extents.

Plains and desert areas have been degraded to a lesser extent. Only 10-20% of the world's drylands, which include temperate grasslands, savannas, and shrublands,scrub, and deciduous forests, have been somewhat degraded. But included in that 10-20% of land is the approximately 9 million square kilometers of seasonally dry-lands that humans have converted to deserts through the process of desertification. The tallgrass prairies of North America, on the other hand, have less than 3% of natural habitat remaining that has not been converted to farmland. Wetlands and marine areas have endured high levels of habitat destruction. More than 50% of wetlands in the U.S. have been destroyed in just the last 200 years.Between 60% and 70% of European wetlands have been completely destroyed.About one-fifth (20%) of marine coastal areas have been highly modified by humans. One-fifth of coral reefs have also been destroyed, and another fifth has been severely degraded by overfishing, pollution, and invasive species; 90% of the Philippines' coral reefs alone have been destroyed. Finally, over 35%mangrove ecosystems worldwide have been destroyed.

Natural causes

Habitat destruction through natural processes such as volcanism, fire, and climate change is well documented in the fossil record. One study shows that habitat fragmentation of tropical rainforests in Euramerica 300 million years ago led to a great loss of amphibian diversity, but simultaneously the drier climate spurred on a burst of diversity among reptiles.

Human causes

Habitat destruction caused by humans includes conversion of land to agriculture,urban sprawl, infrastructure development, and other anthropogenic changes to the characteristics of land. Habitat degradation, fragmentation, and pollution are aspects of habitat destruction caused by humans that do not necessarily involve overt destruction of habitat, yet result in habitat collapse. Desertification,deforestation, and coral reef degradation are specific types of habitat destruction for those areas (deserts, forests, coral reefs).

Geist and Lambin (2002) assessed 152 case studies of net losses of tropical forest cover to determine any patterns in the proximate and underlying causes of tropical deforestation. Their results, yielded as percentages of the case studies in which each parameter was a significant factor, provide a quantitative prioritization of which proximate and underlying causes were the most significant. The proximate causes were clustered into broad categories of agricultural expansion (96%), infrastructure expansion (72%), and wood extraction (67%). Therefore, according to this study, forest conversion to agriculture is the main land use change responsible for tropical deforestation. The specific categories reveal further insight into the specific causes of tropical deforestation: transport extension (64%), commercial wood extraction (52%), permanent cultivation (48%), cattle ranching (46%), shifting (slash and burn) cultivation (41%),subsistence agriculture (40%), and fuel wood extraction for domestic use (28%). One result is that shifting cultivation is not the primary cause of deforestation in all world regions, while transport extension (including the construction of new roads) is the largest single proximate factor responsible for deforestation.

Drivers

While the above-mentioned activities are the proximal or direct causes of habitat destruction in that they actually destroy habitat, this still does not identify why humans destroy habitat. The forces that cause humans to destroy habitat are known as *drivers* of habitat destruction. Demographic, economic, sociopolitical, scientific and technological, and cultural drivers all contribute to habitat destruction.

Demographic drivers include the expanding human population; rate of population increase over time; spatial distribution of people in a given area (urban versus rural), ecosystem type, and country; and the combined effects of poverty, age, family planning, gender, and education status of people in certain areas. Most of the exponential human population growth worldwide is occurring in or close to biodiversity hotspots. This may explain why human population density accounts for 87.9% of the variation in numbers of threatened species across 114 countries, providing indisputable evidence that people play the largest role in decreasing biodiversity. The boom in human population and migration of people into such species-rich regions are making conservation efforts not only more urgent but also more likely to conflict with local human interests. The high local population density in such areas is directly correlated to the poverty status of the local people, most of whom lacking an education and family planning.

From the Geist and Lambin (2002) study described in the previous section, the underlying driving forces were prioritized as follows (with the percent of the 152 cases the factor played a significant role in): economic factors (81%), institutional or policy factors (78%), technological factors (70%), cultural or socio-political factors (66%), and demographic factors (61%). The main economic factors included commercialization and growth of timber markets (68%), which are driven by national and international demands; urban industrial growth (38%); low domestic costs for land, labor, fuel, and timber (32%); and increases in product prices mainly for cash crops (25%). Institutional and policy factors included formal pro-deforestation policies on land development (40%), economic growth including colonization and infrastructure improvement (34%), and subsidies for land-based activities (26%); property rights and land-tenure insecurity (44%); and policy failures such as corruption, lawlessness, or mismanagement (42%). The main technological factor was the poor application of technology in the wood industry (45%), which leads to wasteful logging practices. Within the broad category of cultural and sociopolitical factors are public attitudes and values (63%), individual/household behavior (53%), public unconcern toward forest environments (43%), missing basic values (36%), and unconcern by individuals (32%). Demographic factors were the in-migration of colonizing settlers into sparsely populated forest areas (38%) and growing population density — a result of the first factor — in those areas (25%).

There are also feedbacks and interactions among the proximate and underlying causes of deforestation that can amplify the process. Road construction has the largest feedback effect, because it interacts with—and leads to—the establishment of new settlements and more people, which causes a growth in wood (logging) and food markets. Growth in these markets, in turn, progresses the commercialization of agriculture and logging industries. When these industries become commercialized, they must become more efficient by utilizing larger or more modern machinery that often are worse on the habitat than traditional farming and logging methods. Either way, more land is cleared more rapidly for commercial markets. This common feedback example manifests just how closely related the proximate and underlying causes are to each other.

IMPACT ON HUMAN POPULATION

Habitat destruction vastly increases an area's vulnerability to natural disasters likeflood and drought, crop failure, spread of disease, and water contamination. On the other hand, a healthy ecosystem with good management practices will reduce the chance of these events happening, or will at least mitigate adverse impacts.

Agricultural land can actually suffer from the destruction of the surrounding landscape. Over the past 50 years, the destruction of habitat surrounding agricultural land has degraded approximately 40% of agricultural land worldwide viaerosion, salinization, compaction, nutrient depletion, pollution, and urbanization. Humans also lose direct uses of natural habitat when habitat is destroyed. Aesthetic uses such as bird watching, recreational uses like hunting and fishing, and ecotourism usually rely upon virtually undisturbed habitat. Many people value the complexity of the natural world and are disturbed by the loss of natural habitats and animal or plant species worldwide.

Probably the most profound impact that habitat destruction has on people is the loss of many valuable ecosystem services. Habitat destruction has altered nitrogen, phosphorus, sulfur, and carbon cycles, which has increased the frequency and severity of acid rain, algal blooms, and fish kills in rivers and oceans and contributed tremendously to global climate change. One ecosystem service whose significance is becoming more realized is climate regulation. On a local scale, trees provide windbreaks and shade; on a regional scale, plant transpiration recycles rainwater and maintains constant annual rainfall; on a global scale, plants (especially trees from tropical rainforests) from around the world counter the accumulation of greenhouse gases in the atmosphere by sequestering carbon dioxide through photosynthesis. Other ecosystem services that are diminished or lost altogether as a result of habitat destruction include watershed management, nitrogen fixation, oxygen production, pollination, waste treatment (i.e., the breaking down and immobilization of toxic pollutants), and nutrient recycling of sewage or agricultural runoff.

The loss of trees from the tropical rainforests alone represents a substantial diminishing of the earth's ability to produce oxygen and use up carbon dioxide. These services are becoming even more important as increasing carbon dioxide levels is one of the main contributors to global climate change.

The loss of biodiversity may not directly affect humans, but the indirect effects of losing many species as well as the diversity of ecosystems in general are enormous. When biodiversity is lost, the environment loses many species that provide valuable and unique roles to the ecosystem. The environment and all its inhabitants rely on biodiversity to recover from extreme environmental conditions. When too much biodiversity is lost, a catastrophic event such as an earthquake, flood, or volcanic eruption could cause an ecosystem to crash, and humans would obviously suffer from that. Loss of biodiversity also means that humans are losing animals that could have served as biological control agents and plants that could potentially provide higher-yielding crop varieties, pharmaceutical drugs to cure existing or future diseases or cancer, and new resistant crop varieties for agricultural species susceptible to pesticide-resistant insects or virulent strains of fungi,viruses, and bacteria.

The negative effects of habitat destruction usually impact rural populations more directly than urban populations. Across the globe, poor people suffer the most when natural habitat is destroyed, because less natural habitat means less natural resources per capita, yet wealthier people and countries simply have to pay more to continue to receive more than their per capita share of natural resources.

Another way to view the negative effects of habitat destruction is to look at the opportunity cost of keeping an area undisturbed. In other words, what are people losing out on by taking away a given habitat? A country may increase its food supply by converting forest land to row-crop agriculture, but the value of the same land may be much larger when it can supply natural resources or services such as clean water, timber, ecotourism, or flood regulation and drought control.

Outlook

The rapid expansion of the global human population is increasing the world's food requirement substantially. Simple logic instructs that more people will require more food. In fact, as the world's population increases dramatically, agricultural output will need to increase by at least 50%, over the next 30 years. In the past, continually moving to new land and soils provided a boost in food production to appease the global food demand. That easy fix will no longer be available, however, as more than 98% of all land suitable for agriculture is already in use or degraded beyond repair.

The impending global food crisis will be a major source of habitat destruction. Commercial farmers are going to become desperate to produce more food from the same amount of land, so they will use more fertilizers and less concern for the environment to meet the market demand. Others will seek out new land or will convert other land-uses to agriculture. Agricultural intensification will become widespread at the cost of the environment and its inhabitants. Species will be pushed out of their habitat either directly by habitat destruction or indirectly by fragmentation, degradation, or pollution. Any efforts to protect the world's remaining natural habitat and biodiversity will compete directly with humans' growing demand for natural resources, especially new agricultural lands.

Solutions

In most cases of tropical deforestation, three to four underlying causes are driving two to three proximate causes. This means that a universal policy for controlling tropical deforestation would not be able to address the unique combination of proximate and underlying causes of deforestation in each country. Before any local, national, or international deforestation policies are written and enforced, governmental leaders must acquire a detailed understanding of the complex combination of proximate causes and underlying driving forces of deforestation in a given area or country. This concept, along with many other results about tropical deforestation from the Geist and Lambin study, can easily be applied to habitat destruction in general. Governmental leaders need to take action by addressing the underlying driving forces, rather than merely regulating the proximate causes. In a broader sense, governmental bodies at a local, national, and international scale need to emphasize the following:

1. Considering the many irreplaceable ecosystem services provided by natural habitats.
2. Protecting remaining intact sections of natural habitat.
3. Educating the public about the importance of natural habitat and biodiversity.
4. Developing family planning programs in areas of rapid population growth.
5. Finding ecological ways to increase agricultural output without increasing the total land in production.
6. Preserving habitat corridors to minimize prior damage from fragmented habitats.
7. Reduce human population and expansion.

Habitat conservation

Habitat conservation is a land management practice that seeks to conserve, protect and restore habitat areas for wild plants and animals, especially conservation reliant species, and prevent their extinction, fragmentation or reduction in range. It is a priority of many groups that cannot be easily characterized in terms of any one ideology.

History of the conservation movement

For much of human history, nature had been seen as a resource, one that could be controlled and used for personal and economic gain. The idea was that plants only existed to feed animals and animals only existed to feed man. The land itself had limited value only extending to the resources it could provide such as minerals and oil.

Throughout the 18th and 19th centuries social views started to change and scientific conservation principles were first practically applied to the forests of British India. The conservation ethic that began to evolve included three core principles: that human activity damaged the environment, that there was a civic duty to maintain the environment for future generations, and that scientific, empirically based methods should be applied to ensure this duty was carried out. Sir James Ranald Martin was prominent in promoting this ideology, publishing many medico-topographical reports that demonstrated the scale of damage wrought through large-scale deforestation and desiccation, and lobbying extensively for the institutionalization of forest conservation activities in British India through the establishment of Forest Departments.

The Madras Board of Revenue started local conservation efforts in 1842, headed by Alexander Gibson, a professional-botanist who systematically adopted a forest conservation program based on scientific principles. This was the first case of state conservation management of forests in the world. Governor-General Lord Dalhousie introduced the first permanent and large-scale forest conservation program in the world in 1855, a model that soon spread to other colonies, as well the United States, where Yellowstone National Park was opened in 1872 as the world's first national park.

Rather than focusing on the economic or material benefits associated with nature, humans began to appreciate the value of nature itself and the need to protect pristine wilderness. By the middle of the 20th century countries such as the United States, Canada, and Britain understood this appreciation and instigated laws and legislation in order to ensure that the most fragile and beautiful environments would be protected for generations to come. Today with the help of NGO's, not-for profit organizations and governments world-wide there is a stronger movement taking place, with a deeper understanding of habitat conservation with the aim of protecting delicate habitats and preserving biodiversity on a global scale. The commitment and actions of small volunteering association in villages and towns, that endeavour to emulate the work done by well known Conservation Organisations, is paramount in ensuring generations that follow understand the importance of conserving natural resources. A village conservation group with the mission statement "We are committed to protecting and enhancing the natural environment in and around the adjoining villages of Ouston and Urpeth." may one day inspire a child who becomes the employee of a worldwide conservation organisation.

Values of natural habitat

The natural environment is a source for a wide range of resources that can be exploited for economic profit, for example timber is harvested from forests and clean water is obtained from natural streams. However, land development from anthropogenic economic growth often causes a decline in the ecological integrity of nearby natural habitat. For instance, this was an issue in the northern rocky mountains of the USA.

However, there is also economic value in conserving natural habitat. Financial profit can be made from tourist revenue, particularly in the tropics where species diversity is high. The cost of repairing damaged ecosystems is considered to be much higher than the cost of conserving natural ecosystems.

Measuring the worth of conserving different habitat areas is often criticized as being too utilitarian from a philosophical point of view.

Biodiversity

Habitat conservation is important in maintaining biodiversity, an essential part of global food security. There is evidence to support a trend of accelerating erosion of the genetic resources of agricultural plants and animals. An increase in genetic similarity of agricultural plants and animals means an increased risk of food loss from major epidemics. Wild species of agricultural plants have been found to be more resistant to disease, for example the wild corn species Teosinte is resistant to 4 corn diseases that affect human grown crops. A combination of seed banking and habitat conservation has been proposed to maintain plant diversity for food security purposes.

Classifying environmental values

Pearce and Moran outlined the following method for classifying environmental uses.

- Direct extractive uses: e.g. timber from forests, food from plants and animals
- Indirect uses: e.g. ecosystem services like flood control, pest control, erosion protection
- Optional uses: future possibilities e.g. unknown but potential use of plants in chemistry/medicine
- Non-use values:

- Bequest value (benefit of an individual who knows that others may benefit from it in future)
- Passive use value (sympathy for natural environment, enjoyment of the mere existence of a particular species)

Impacts

Natural causes

Habitat loss and destruction can occur both naturally and through anthropogenic causes. Events leading to natural habitat loss include climate change, catastrophic events such as volcanic explosions and through the interactions of invasive and non-invasive species. Natural climate change, events have previously been the cause of many widespread and large scale losses in habitat. For example, some of the mass extinction events generally referred to as the "Big Five" have coincided with large scale such as the Earth entering an ice age, or alternate warming events. Other events in the big five also have their roots in natural causes, such as volcanic explosions and meteor collisions. The Chicxulub impact is one such example, which has previously caused widespread losses in habitat as the Earth either received less sunlight or grew colder, causing certain fauna and flora to flourish whilst others perished. Previously known warm areas in the tropics, the most sensitive habitats on Earth, grew colder, and areas such as Australia developed radically different flora and fauna to those seen today. The big five mass extinction events have also been linked to sea level changes, indicating that large scale marine species loss was strongly influenced by loss in marine habitats, particularly shelf habitats. Methane-driven oceanic eruptions have also been shown to have caused smaller mass extinction events.

Human impacts

Humans have been the cause of many species' extinction. Due to humans' changing and modifying their environment, the habitat of other species often become altered or destroyed as a result of human actions. Even before the modern industrial era, humans were having widespread, and major effects on the environment. A good example of this is found in Aboriginal Australians and Australian megafauna. Aboriginal hunting practices, which included burning large sections of forest at a time, eventually altered and changed Australia's vegetation so much that many herbivorous megafauna species were left with no habitat and were driven into extinction. Once herbivorous megafauna species became extinct, carnivorous megafauna species soon followed. In the recent past, humans have been responsible for causing more extinctions within a given period of time than ever before. Deforestation, pollution, anthropogenic climate change and human settlements have all been driving forces in altering or destroying habitats. The destruction of ecosystems such as rainforests has resulted in countless habitats being destroyed. These biodiversity hotspots are home to millions of habitat specialists, which do not exist beyond a tiny area. Once their habitat is destroyed, they cease to exist.This destruction has a follow-on effect, as species which coexist or depend upon the existence of other species also become extinct, eventually resulting in the collapse of an entire ecosystem. These time-delayed extinctions are referred to as the extinction debt, which is the result of destroying and fragmenting habitats. As a result of anthropogenic modification of the environment, the extinction rate has climbed to the point where the Earth is now within a sixth mass extinction event, as commonly agreed by biologists. This has been particularly evident, for example, in the rapid decline in the number of amphibian species worldwide.

APPROACHES AND METHODS OF HABITAT CONSERVATION

Determining the size, type and location of habitat to conserve is a complex area of conservation biology. Although difficult to measure and predict, the conservation value of a habitat is often a reflection of the quality (e.g. species abundance and diversity), endangerment of encompassing ecosystems, and spatial distribution of that habitat.

Identifying priority habitats for conservation

Habitat conservation is vital for protecting species and ecological processes. It is important to conserve and protect the space/ area in which that species occupies. Therefore, areas classified as 'biodiversity hotspots', or those in which a flagship, umbrella, or endangered species inhabits are often the habitats that are given precedence over others. Species that possess an elevated risk of extinction are given the highest priority and as a result of conserving their habitat, other species in that community are protected thus serving as an element of gap analysis. In the United States of America, aHabitat Conservation Plan (HCP) is often developed to conserve the environment in which a specific species inhabits. Under the U.S. Endangered Species Act (ESA) the habitat that requires protection in an HCP is referred to as the 'critical habitat'. Multiple-species HCPs are becoming more favourable than single-species HCPs as they can potentially protect an array of species before they warrant listing under the ESA, as well as being able to conserve broad ecosystem components and processes. As of January 2007, 484 HCPs were permitted across the United States, 40 of which covered 10 or more species. The San Diego Multiple Species Conservation Plan (MSCP) encompasses 85 species in a total area of 26,000-km2. Its aim is to protect the habitats of multiple species and overall biodiversity by minimizing development in sensitive areas.

HCPs require clearly defined goals and objectives, efficient monitoring programs, as well as successful communication and collaboration with stakeholders and land owners in the area. Reserve design is also important and requires a high level of planning and management in order to achieve the goals of the HCP. Successful reserve design often takes the form of a hierarchical system with the most valued habitats requiring high protection being surrounded by buffer habitats that have a lower protection status. Like HCPs, hierarchical reserve design is a method most often used to protect a single species, and as a result habitat corridors are maintained, edge effects are reduced and a broader suite of species are protected.

How much habitat is enough

A range of methods and models currently exist that can be used to determine how much habitat is to be conserved in order to sustain a viable population. Modelling tools often rely on the spatial scale of the area as an indicator of conservation value. There has been an increase in emphasis on conserving few large areas of habitat as opposed to many small areas. This idea is often referred to as the "single large or several small", SLOSS debate, and is a highly controversial area among conservation biologists and ecologists. The reasons behind the argument that "larger is better" include the reduction in the negative impacts of patch edge effects, the general idea that species richness increases with habitat area and the ability of larger habitats to support greater populations with lower extinction probabilities. Noss & Cooperrider support the "larger is better" claim and developed a model that implies areas of habitat less than 1000ha are "tiny" and of low conservation value. However, Shwartz suggests that although "larger is better", this does not imply that "small is bad". Shwartz argues that human induced habitat loss leaves no alternative to conserving small areas. Furthermore, he suggests many endangered species which are of high conservation value, may only be restricted to small isolated patches of habitat, and thus would be overlooked if larger areas were given a higher priority. The shift to conserving larger areas is somewhat justified in society by placing more value on larger vertebrate species, which naturally have larger habitat requirements.

Examples of current conservation organizations

The Nature Conservancy

Since its formation in 1951 The Nature Conservancy has slowly developed into one of the world's largest conservation organizations. Currently operating in over 30 countries, across 5 continents world-wide, The Nature Conservancy aims to protect nature and its assets for future generations. The organization purchases land or accepts land donations with the intension of conserving its natural resources. In 1955 The Nature Conservancy purchased its first 60-acre plot near the New York/Connecticut border in the United States of America. Today the Conservancy has expanded to protect over

119 million acres of land, 5,000 river miles as well as participating in over 1000 marine protection programs across the globe. Since its beginnings The Nature Conservancy has understood the benefit in taking a scientific approach towards habitat conservation. For the last decade the organization has been using a collaborative, scientific method known as 'Conservation by Design'. By collecting and analyzing scientific data The Conservancy is able to holistically approach the protection of various ecosystems. This process determines the habitats that need protection, specific elements that should be conserved as well as monitoring progress so more efficient practices can be developed for the future.

The Nature Conservancy currently has a large number of diverse projects in operation. They work with countries around the world to protect forests, river systems, oceans, deserts and grasslands. In all cases the aim is to provide a sustainable environment for both the plant and animal life forms that depend on them as well as all future generations to come.turtles

World Wildlife Fund (WWF)

The World Wildlife Fund (WWF) was first formed in after a group of passionate conservationists signed what is now referred to as the Morges Manifesto., WWF is currently operating in over 100 countries across 5 continents with a current listing of over 5 million supporters. One of the first projects of WWF was assisting in the creation of the Charles Darwin Research Foundation which aided in the protection of diverse range of unique species existing on the Galápagos' Islands, Ecuador. It was also a WWF grant that helped with the formation of the College of African Wildlife Management in Tanzania which today focuses on teaching a wide range of protected area management skills in areas such as ecology, range management and law enforcement. The WWF has since gone on to aid in the protection of land in Spain, creating the Coto Doñana National Park in order to conserve migratory birds and The Democratic Republic of Congo, home to the world's largest protected wetlands. The WWF also initiated a debt-for-nature concept which allows the country to put funds normally allocated to paying off national debt, into conservation programs that protect its natural landscapes. Countries currently participating include Madagascar, the first country to participate which since 1989 has generated over $US50 million towards preservation, Bolivia, Costa Rica, Ecuador, Gabon, the Philippines and Zambia.

Rare Conservation

Rare has been in operation since 1973 with current global partners in over 50 countries and offices in the United States of America, Mexico, the

Philippines, China and Indonesia. Rare focuses on the human activity that threatens biodiversity and habitats such as overfishing and unsustainable agriculture. By engaging local communities and changing behaviour Rare has been able to launch campaigns to protect areas in most need of conservation. The key aspect of Rare's methodology is their "Pride Campaign's". For example, in the Andes in South America, Rare has partnered with 11 different sites with the intention of creating incentives to develop watershed protection practices. In the Southeast Asia's "coral triangle" Rare is training fishers in local communities to better manage the areas around the coral reefs in order to lessen human impact. Such programs last for three years with the aim of changing community attitudes so as to conserve fragile habitats and provide ecological protection for years to come.

WWF Netherlands

WWF Netherlands, along with ARK Nature, Wild Wonders of Europe and Conservation Capital have started the Rewilding Europe project. This project intents to rewild several areas in Europe.

Biotope

Biotope is an area of uniform environmental conditions providing a living place for a specific assemblage of plants and animals. Biotope is almost synonymous with the term habitat, which is more commonly used in English-speaking countries. However, in some countries these two terms are distinguished: the subject of a habitat is a species or a population, the subject of a biotope is a biological community.

It is an English loanword derived from the German "*Biotop*", which in turn came from the Greek bios='life' or 'organism' and topos='place'. (The related word *geotope* has made its way into the English language by the same route, from the German "*Geotop*".) The word biotope, literally translated, means an area where life lives.

ECOLOGY

The concept of a biotope was first advocated by Ernst Haeckel (1834-1919), a German zoologist famous for therecapitulation theory. In his book *General Morphology* (1866), which defines the term "ecology", he stresses the importance of the concept of habitat as a prerequisite for an organism's existence. Heackel also explains that with one ecosystem, itsbiota is shaped by environmental factors (such as water, soil, and geographical features) and interaction among living things; the original idea of a biotope was closely related to evolutional theory. Following this, F. Dahl, a professor at the Berlin Zoological Museum, referred to this ecological system as a "biotope" (biotop) (1908).

Biotope restoration

Although the term "biotope" is considered to be a technical word with respect to ecology, in recent years the term is more generally used in administrative and civic activities. Since the 1970s the term "biotope" has received great attention as a keyword throughout Europe (mainly Germany) for the preservation, regeneration, and creation of natural environmental settings. Used in this context, the term "biotope" often refers to a smaller and more specific ecology and is very familiar to human life.

In Germany especially, activities related to regenerating biotopes are enthusiastically received. These activities include—

- Making roof gardens,
- Reconstructing rivers to restore their natural qualities,
- Leaving bushes or trees on farms,
- Building nature parks along motorways "Autobahn,"
- Making school gardens or ponds by considering the ecosystem, and
- Bearing in mind ecological considerations in private gardens.

Various sectors play a part in these activities, including architecture, civil engineering, urban planning, traffic, agriculture, river engineering, limnology, biology, education, landscape gardening, and domestic gardening. In all fields, all sorts of people are seeking a viable way for humans to respect other living things. The term "biotope" would include a complete environmental approach.

Characteristics

The following four points are the chief characteristics of biotopes.

Microscale

A biotope is generally not considered to be a large-scale phenomenon. For example, a biotope might be a neighbouring park, a back garden, even potted plants or a fish tank on a porch. In other words, the biotope is not a macroscopic but a microscopic approach to preserving the ecosystem and biological diversity. So biotopes fit into ordinary people's daily activities and lives, with more people being able to take part in biotope creation and continuing management.

Biotope networks

It is commonly emphasised that biotopes should not be isolated. Instead biotopes need to be connected to each other and other surrounding life for without these connections to life-forms such as animals and plants, biotopes would not effectively work as a place in which diverse organisms live. So one of the most effective strategies for regenerating biotopes is to plan a *stretch* of biotopes, not just a *point* where animals and plants come and go. (Such an organic traffic course is called acorridor.) In the stretch method, the centre of the network would be large green tracts of land: a forest, natural park, or cemetery. By connecting parcels of land with smaller biotope areas such as a green belt along the river, small town parks, gardens, or even roadside trees, biotopes can exist in a network. In other words, a biotope is an open not a closed system and is a practicable strategy.

Human daily life

The term "biotope" does not apply to biosphere reserves, which are completely separate from humans and become the object of human admiration. Instead, it is an active part of human daily life. For example, an ornamental flower bed may be considered a biotope (albeit rather a small one) since it enhances the experience of daily life. An area that has many functions, such as human living space, and is home to other living things, whether plant or animal, can be considered a biosphere reserve.

Artificial

When artificial items are introduced to a biotope setting, their design and arrangement is of great importance for biotope regeneration. Tree-planting areas where the surface is uneven results in plants that sprout and the nesting of small insects. A mat or net made from natural fibres will gradually biodegrade as it is exposed to the weather. So there is no binomial opposition between the natural and the artificial in a biotope. Rather, such artificial materials are widely used.

Germany

It is especially characteristic in Germany, which is the birthplace of the term biotope, that the authorities take the initiative in conserving biotopes, maintaining consistency with urban or rural planning and considering the regions' history and landscape.

Legal basis

The federal nature protection law (Bundesnaturs-chutzgesetz, since 1976) requires that wild animals and plants and their community should be protected as part of the ecosystem in the specific diversity that has grown naturally and historically, and their biotope and other living conditions should be protected, preserved, developed, and restored. (Number 9, Clause 1, Article 2). The law also requires that some kinds of biotope that are full of a specific variety should not be harmed by development. So there is a law that mandates the protection of biotopes. There is also a provincial law corresponding to the federal one. Such developments were uncommon in those times.

Landscape plan

Many German states are obliged by law to produce a landscape plan (Landschaftsplan) as part of their urban planning, though these plans vary somewhat from place to place. The purpose of the "Landschaftsplan" is to protect the region's environment and landscape. These plans use text and figures to describe the present environmental state and proposed remedies. They consider, for example, the regional lie of the land, climate, wind direction, soil, ground water, type of biotope, distribution of animals and plants, inhabitants' welfare and competition with development projects.

Citizen welfare

Biotope preservation in cities also emphasises recreation and relaxation for citizens and improving the urban environment. For example, in the reserve of Karlsruhe in Baden-Württemberg people can cycle on the bike path or walk the dog, although it is forbidden to gather plants and animals there or walk in the exclusion zone. At the core of biotope preservation is the idea that if civic life is surrounded by a rich profusion of nature whose background is in local history and culture, it is improved by protecting nature and preserving the landscape.

Aquaria

The term "biotope" is also used by aquarium hobbyists to describe an aquarium setup that tries to simulate the naturalhabitat of specific fish. The idea is to replicate conditions such as water temperature, natural plants, water type (fresh,saline or brackish), lighting, and other native fish, which represent a particular real-world biotope. An example of one South American biotope type might be lots of bog-wood, a few native plants, dark substrata and subdued lighting with floating plants, along with marbled hatchets, angels, cardinals, otos, corys and plecostomus.

ECOLOGICAL NICHE

In ecology, a niche is a term with a variety of meanings related to the behavior of a species living under specific environmental conditions. The ecological niche describes how an organism or population responds to the distribution of resources and competitors (for example, by growing when resources are abundant, and when predators, parasites and pathogens are scarce) and how it in turn alters those same factors (for example, limiting access to resources by other organisms, acting as a food source for predators and a consumer of prey). "The type and number of variables comprising the dimensions of an environmental niche vary from one species to another [and] the relative importance of particular environmental variables for a species may vary according to the geographic and biotic contexts".

The notion of ecological niche is central to ecological biogeography, which focuses on spatial patterns of ecological communities. "Species distributions and their dynamics over time result from properties of the species, environmental variation..., and interactions between the two — in particular the abilities of some species, especially our own, to modify their environments and alter the range dynamics of many other species." Alteration of an ecological niche by its inhabitants is the topic of niche construction.

The majority of species exist in a standard ecological niche, but there are exceptions. A premier example of a non-standard niche filling species is the flightless, ground-dwelling kiwi bird of New Zealand, which feeds on worms and other ground creatures, and lives its life in a mammal niche. Island biogeography can help explain island species and associated unfilled niches.

Grinnellian niche

The ecological meaning of niche comes from the meaning of niche as a recess in a wall for a statue, which itself is probably derived from the Middle French word *nicher*, meaning *to nest*. The term was coined by the naturalist Joseph Grinnell in 1917, in his paper "The niche relationships of the California Thrasher". The Grinnellian niche concept embodies the idea that the niche of a species is determined by the habitat in which it lives and its accompanying behavioral adaptations. In other words, the niche is the sum of the habitat requirements and behaviors that allow a species to persist and produce offspring. For example, the behavior of the California Thrasher is consistent with the chaparral habitat it lives in—it breeds and feeds in the underbrush and escapes from its predators by shuffling from underbrush to underbrush. Its 'niche' is defined by the felicitous complementing of the thrasher's behavior and physical traits (camouflaging color, short wings, strong legs) with this habitat.

This perspective of niche allows for the existence of both ecological equivalents and empty niches. An ecological equivalent to an organism is an organism from a different taxonomic group exhibiting similar adaptations in a similar habitat, an example being the different succulents found in American and African deserts, cactus and euphorbia. As another example, the *Anolis* lizards of the Greater Antilles are a rare example of convergent evolution, adaptive radiation, and the existence of ecological equivalents: the *Anolis* lizards evolved in similar microhabitats independently of each other and resulted in the same ecomorphs across all four islands.

Eltonian niche

In 1927 Charles Sutherland Elton, a British ecologist, defined a niche as follows: "The 'niche' of an animal means its place in the biotic environment, *its relations to food and enemies.*"

Elton classified niches according to foraging activities ("food habits"): "For instance there is the niche that is filled by birds of prey which eat small animals such as shrews and mice. In an oak wood this niche is filled by tawny owls, while in the open grassland it is occupied by kestrels. The existence of this carnivore niche is dependent on the further fact that mice form a definite herbivore niche in many different associations, although the actual species of mice may be quite different."

Hutchinsonian niche

The Hutchinsonian niche is an n-dimensional hypervolume, where the dimensions are environmental conditions and resources, that define the requirements of an individual or a species to practice "its" way of life, more particularly, for its population to persist. The "hypervolume" defines the multi-dimensional space of resources (e.g., light, nutrients, structure, etc.) available to (and specifically used by) organisms, and "all species other than those under consideration are regarded as part of the coordinate system."

The niche concept was popularized by the zoologist G. Evelyn Hutchinson in 1957. Hutchinson wanted to know why there are so many types of organisms in any one habitat. His work inspired many others to develop models to explain how many and how similar coexisting species could be within a given community, and led to the concepts of 'niche breadth' (the variety of resources or habitats used by a given species), 'niche partitioning' (resource differentiation by coexisting species), and 'niche overlap' (overlap of resource use by different species).

Statistics were introduced into the Hutchinson niche by Robert MacArthur andRichard Levins using the 'resource-utilization' niche employing histograms to describe the 'frequency of occurrence' as a function of a Hutchinson coordinate. So, for instance, a Gaussian might describe the frequency with which a species ate prey of a certain size, giving a more detailed niche description than simply specifying some median or average prey size. One advantage in using statistics is illustrated in the figure, where it is clear that for the narrower distributions (top) there is no competition for prey between the extreme left and extreme right species, while for the broader distribution (bottom), niche overlap indicates competition exists between all species. For such a bell-shaped distribution, the *position, width* and *form* of the niche correspond to the *mean, standard deviation* and the actual distribution itself.

An organism free of interference from other species could use the full range of conditions (biotic and abiotic) and resources in which it could survive and reproduce which is called its fundamental niche. However, as a result of pressure from, and interactions with, other organisms (i.e. inter-specific competition) species are usually forced to occupy a niche that is narrower than this, and to which they are mostly highly adapted. This is termed the realized niche. Hutchinson used the idea of competition for resources as the primary mechanism driving ecology, but overemphasis upon this focus has proved to be a handicap for the niche concept. In particular, overemphasis upon a species' dependence upon resources has led to too little emphasis upon the effects of organisms on their environment, for instance, colonization and invasions.

The term adaptive zone was coined by the paleontologist George Gaylord Simpson to explain how a population could jump from one niche to another that suited it, jump to an 'adaptive zone', made available by virtue of some modification, or possibly a change in the food chain, that made the adaptive zone available to it without a discontinuity in its way of life because the group was 'pre-adapted' to the new ecological opportunity.

Hutchinson's "niche" (a description of the ecological space occupied by a species) is subtly different from the "niche" as defined by Grinnell (an ecological role, that may or may not be actually filled by a species—see vacant niches).

A niche is a very specific segment of ecospace occupied by a single species. On the presumption that no two species are identical in all respects (called Hardin's 'axiom of inequality') and the competitive exclusion principle, *some* resource or adaptive dimension will provide a niche specific to each species. Species can however share a 'mode of life' or 'autecological strategy' which are broader definitions of ecospace. For example, Australian grasslands species, though different from those of the Great Plains grasslands, exhibit similar modes of life.

Once a niche is left vacant, other organisms can fill that position. For example, the niche that was left vacant by the extinction of the tarpan has been filled by other animals (in particular a small horse breed, the konik). Also, when plants and animals are introduced into a new environment, they have the potential to occupy or invade the niche or niches of native organisms, often outcompeting the indigenous species. Introduction of non-indigenous species to non-native habitats by humans often results in biological pollution by the exotic or invasive species.

The mathematical representation of a species' fundamental niche in ecological space, and its subsequent projection back into geographic space, is the domain ofniche modelling.

Parameters

The different dimensions, or plot axes, of a niche represent different biotic and abiotic variables. These factors may include descriptions of the organism's life history, habitat, trophic position (place in the food chain), and geographic range. According to the competitive exclusion principle, no two species can occupy the same niche in the same environment for a long time. The parameters of a realized niche are described by the realized niche width of that species.

Niche construction

Niche construction is the process in which an organism alters its own (or other species') environment, often but not always in a manner that increases its chances of survival. Changes that organisms bring about in their worlds that are of no evolutionary or ecological consequence are not examples of niche construction. Several biologists have argued that niche construction is as important to evolution as natural selection (i.e., not only does an environment cause changes in species through selection, but species also cause changes in their environment through niche construction). This back-and-forth creates a feedback relationship between natural selection and niche-construction: when organisms affect their environment, that change can then cause a shift in what traits are being naturally selected for. The effect of niche construction is especially pronounced in situations where environmental alterations persist for several generations, introducing the evolutionary role of ecological inheritance. Less drastic niche-constructing behaviors are also quite possible for an organism. This theory, in conjunction with natural selection, shows that organisms inherit two legacies from their ancestors: genes and a modified environment. Together, these two evolutionary mechanisms determine a population's fitness and what adaptations those organisms develop in the continuation for their survival.

Examples

- Earthworms: Through a process of modification, earthworms chemically alter the soil in which they live. This change in soil chemistry stimulates an increased fitness in earthworm populations. The subsequent chemical composition of soil produced by earthworm activity also benefits the growth of proximal species of plants and other biota present in the soil.
- Lemon Ants (*Myrmelachista schumanni*): This species of ants employs a specialized method of suppression that regulates the growth of certain trees. Lemon ants make their homes in the bodies of *Duroia hirsuta*, a species of tree found in the Amazonian rainforest of Peru. To ensure the prevalence of *Duroia* trees, Lemon ants employ self-derived quantities of formic acid (a chemical fairly common among species of ants) as a tenacious herbicide.

The end result alters the composition of their forest habitat drastically by wiping out arboreal species ill-suited for colony habitation. When observed by humans, the subsequent altered ecologies perpetuated by these ants have been termed the Devil's Gardens.

- Beavers: In the construction of their dams, beavers drastically shape and alter the ecosystem in which they live. Deforestation, effects on soil structure, root structure, turbidity of water, allocation of water and the supply of water downstream are just a handful of exemplars defining beaver niche construction. Beavers express a clear example of the diverse effects perpetuated by the construction of a niche. In the mammalian kingdom beavers are one of the greatest proximal modifiers.
- Diatoms in the Bay of Fundy, Canada, provide another example of an ecosystem engineer. Benthic diatoms living inestuarine sediments secrete carbohydrate exudates that bind the sand and stabilize the environment. The diatoms cause a physical state change in the properties of the sand that allows other organisms to colonize the area. The concept of ecosystem engineering brings new conceptual implications for the discipline of conservation biology.
- Pine Trees & Chaparrals: Chaparrals and pines have been found to express niche constructing behavior in response to the effects of forest fire. These trees and shrubs increase the frequency of fire by affecting the composition of the forest floor. They achieve this end through the dispersal of needles, cones, seeds and oils, essentially littering the forest floor. The benefit of this activity is compounded by an adaptation in these particular flora that has selected for a resistance to fire. The evolutionary exchange (made between both niche construction and natural selection) allows the fire-resistant pine and chaparral to exploit the chemical change that occurs in soil after organic matter has been burned.
- Humans: See subsection below.

Implications

Niche construction has many implications for the human sciences, more specifically human sociobiology,evolutionary psychology, and human behavioral ecology. Standard evolutionary theory only allows for cultural processes to affect genetic evolution by influencing the individual, and depends on the ability of that individual to survive and pass on its genes to the next generation. Cultural processes are viewed merely as an aspect of the human phenotype and are not believed to be consequential to human evolution. Cultural diversity is believed to reflect variation in the environments that different populations of humans evolved in, and nothing else. This theory overlooks the fact that humans can modify their selective environments through cultural activity, thus feeding back to affect selection. "Cultural processes add a second knowledge inheritance system to the evolutionary process through which socially learned information is accrued, stored, and transmitted between individuals both within and between generations."

With the addition of language to the human culture came an increased mental capacity. This allowed for human adaptation of the environment to be a learned process, unlike nonhuman species, whose adaptive process is instinctual. This resulted in the acceleration of environmental, behavioral and genetic modifications. As niche construction advocate Derek Bickerton writes, "We could construct our niches without having to wait on interminable rounds of feedback between genes and behavior."

A theory on gene-culture coevolution calls for a more integrated relationship between genetic evolution and cultural processes than standard evolutionary theory. In this model, cultural activities are believed to affect the evolutionary process by modifying selection pressures. In other words, cultural change has the capacity to codirect its population's genetic evolution. Mathematical and conceptual models including investigations of language, handedness, the emergence of incest taboos, the coevolution of hereditary deafness and sign language, and sexual selection with a culturally transmitted mating preference demonstrate this theory. However, this theory is still dependent on standard evolutionary theory because it requires that cultural processes only affect genes directly, not allowing for any intermediate factors in the environment to interact with these processes at an evolutionary level. This theory exists on a dual inheritance system consisting solely of genes and cultural activity. "The dual inheritance system is a way to include interactions between nature and nurture in a tractable system." In most cases this theory works smoothly, however there are instances where cultural activities create changes in the abiotic environment that then affect selection pressures.

The speed at which humans are able to construct niches modifies the selection pressures and either genetic evolution or further niche construction can result. An example of genetic evolution through niche construction with the inclusion of an abiotic factor: Yam cultivators in West Africa cut clearings in forests to grow crops, but resulted in much standing water which attracted mosquitoes and increased the rate of malaria. This caused a modification to the selection pressure for the sickle-cell allele that protects against malaria. Evolutionary change is thus furthered. Example of further niche construction: Humans change the environment through pollution. The effects of pollution are alleviated by the innovation and use of a new technology. This cultural response to a constructed niche allows for a change in environment and a lack of change in genetics. Only if a new technology is not created or effective will evolutionary change occur. Humans are able to sustain adaptiveness by responding to ancestral niche construction through further cultural niche construction.

The addition of niche construction to the study of evolutionary processes forces scientists to accept that cultural activity is not the *reason* that humans are able to modify their environments, but is simply their *primary means* of doing the same thing that other species do. The fact that a large number of cultural processes are learned rather than genetically encoded into the individual, makes human culture an incredibly powerful method of niche construction. "Most of the time, cultural processes can be regarded as a shortcut to acquiring adaptive information, as individuals rapidly learn, or are shown, what to eat, where to live, or how to avoid danger by doing what other more knowledgeable individuals do."

CONSEQUENCES

As creatures move into new niches, they can have a significant effect on the world around them. The first consequence that arises from niche construction is that the organisms have changed the environment on which they live. A good example of this is the leafcutter ants mentioned above. Leafcutter colonies can grow to massive sizes and contain millions of individuals. Such a large amount of ants require a large food supply. In order to obtain this, ants need to stock pile a large amount of foliage clippings to feed their crop of fungi. This can devastate surrounding plant life, especially young saplings that need to obtain all the sunlight they can in the rainforests.

Another important consequence is that they can affect natural selection pressures put on a species. The common cuckoo bird is an excellent example of such a consequence. This species of bird parasites other birds by laying their eggs in the other species nest. This had led to several adaptations among the cuckoo's, one of which is a short incubation time for their eggs. The eggs need to hatch first so that the chick can push the other species' eggs out of the nest, ensuring it has no competition for the parent attention. Another adaptation it has acquired is that the chick mimics the calls of multiple young chicks, so that the parents are bringing in food not just for one offspring, but a whole brood.

Niche differentiation

The term niche differentiation (synonymous with niche segregation, niche separation and niche partitioning), as it applies to the field of ecology, refers to the process by which natural selection drives competing species into different patterns of resource use or different niches. This process allows two species to partition certain resources so that one species does not out-compete the other as dictated by the competitive exclusion principle; thus, coexistence is obtained through the differentiation of their ecological niches. Niche partitioning may not occur if there is sufficient geographic and ecological space for organisms to expand into.

Niche differentiation is a process which occurs through several different modes and on multiple temporal and spatial scales. In most cases, niche differentiation has created a relationship between two species where current competition is small or non-existent. Because of this, the presence of niche differentiation can be methodologically difficult to prove or disprove.

The lack of evidence for current or past competition can blur the line between 1) two competitive species differentiating their niches to allow coexistence as opposed to 2) two non-competing species which occupy similar niches. It is important to keep in mind that niche differentiation and inter-specific competition cannot always be considered linked.

As an example of resource partitioning, seven *Anolis* lizards in tropical rainforest share common food needs — mainlyinsects. They avoid competition by occupying different sections of the rainforest. Some live on the leaf litter floor while others live on shady branches, thereby avoiding competition over food in those sections of the forest. All resources are subject to partitioning, for example; space, food, nesting sites. This minimizes competition between similar species.

Niche differentiation: detection and quantification

The Lotka-Volterra equation states that two competing species can coexist when intra-specific (within species) competition is greater than inter-specific (between species) competition (Armstrong and McGehee 1981). Since niche differentiation concentrates competition within-species, due to a decrease in between-species competition, the Lotka-Volterra model predicts that niche differentiation of any degree will result in coexistence.

In reality, this still leaves the question of how much differentiation is needed for coexistence (Hutchinson 1959). A vague answer to this question is that the more similar two species are, the more finely balanced the suitability of their environment must be in order to allow coexistence. There are limits to the amount of niche differentiation required for coexistence, and this can vary with the type of resource, the nature of the environment, and the amount of variation both within and between the species.

To answer questions about niche differentiation, it is necessary for ecologists to be able to detect, measure, and quantify the niches of different coexisting and competing species. This is often done through a combination of detailed natural history studies, controlled experiments (to determine the strength of competition), and mathematical models (Strong 1982, Leibold 1995). To understand the mechanisms of niche differentiation and competition, much data must be gathered on how the two species interact, how they use their resources, and the type of ecosystem in which they exist, among other factors. In addition, several mathematical models exist to quantify niche breadth, competition, and coexistence (Bastolla et al. 2005). However, regardless of methods used, niches and competition can be distinctly difficult to measure quantitatively, and this makes detection and demonstration of niche differentiation difficult and complex.

Development of niche differentiation

Over time, two competing species can either coexist, through niche differentiation or other means, or compete until one species becomes locally extinct. Several theories exist for how niche differentiation arises or evolves given these two possible outcomes.

CURRENT COMPETITION (THE GHOST OF COMPETITION PRESENT)

Niche differentiation can arise from current competition. For instance, species X has a fundamental niche of the entire slope of a hillside, but its realized niche is only the top portion of the slope because species Y, which is a better competitor but cannot survive on the top portion of the slope, has excluded it from the lower portion of the slope. With this scenario, competition will continue indefinitely in the middle of the slope between these two species. Because of this, detection of the presence of niche differentiation (through competition) will be relatively easy.

It is also important to remember that there is no evolutionary change of the individual species in this case; rather this is an ecological effect of species Y out-competing species X within the bounds of species Y's fundamental niche.

Via past extinctions (The Ghost of Competition Past)

Another way by which niche differentiation can arise is via the previous elimination of species without realized niches. This asserts that at some point in the past, several species inhabited an area, and all of these species had overlapping fundamental niches.

However, through competitive exclusion, the less competitive species were eliminated, leaving only the species that were able to coexist (i.e. the most competitive species whose realized niches did not overlap). Again, this process does not include any evolutionary change of individual species, but it is merely the product of the competitive exclusion principle. Also, because no species is out-competing any other species in the final community, the presence of niche differentiation will be difficult or impossible to detect.

Evolving differences

Finally, niche differentiation can arise as an evolutionary effect of competition. In this case, two competing species will evolve different patterns of resource use so as to avoid competition. Here too, current competition is absent or low, and therefore detection of niche differentiation is difficult or impossible.

Types of niche differentiation

Resource partitioning

When two species partition (divide) a resource based on behavioral or morphological variation, it is termed differential resource utilization or resource partitioning. There are three types of differential resource utilization.

Temporal partitioning

Temporal resource partitioning occurs when two species eliminate direct competition by utilizing the same resource at different times. This can be on a daily scale (e.g. one species of spiny mouse feeds on insects during the day while a second species of spiny mouse feeds on the same insects at night, Kronfeld-Schor and Dayan 1999) or on a longer, seasonal scale. An instance of the latter would be reproductive asynchrony, or the division of resources by the separation of breeding periods. An example of reproductive asynchrony would be two competing species of frog offsetting their breeding periods. By doing this the first species' tadpoles will have graduated to a different food resource by the time the tadpoles of the second species are hatching (Lawler and Morin 1993).

Spatial partitioning

Spatial resource partitioning occurs when two competing species use the same resource by occupying different areas or habitats within the range of occurrence of the resource. Spatial partitioning can occur at small scales (microhabitat differentiation) or at large scales (geographical differentiation). Microhabitat differentiation occurs when two competing species with overlapping home ranges partition a resource. Two examples would be different species of fish feeding at different depths in a lake or different species of monkey feeding at different heights in a tree. Geographical differentiation is when two competing species have non-overlapping home ranges and thus partition resources. An example might be given with monkeys again: two competing species of monkey using the same species of fruit trees, but in different areas of the forest.

Morphological differentiation

The final type of differential resource utilization is morphological differentiation or niche complementarity. Morphological differentiation happens when two competing species evolve differing morphologies to allow them to use a resource in different ways. A classic example of this is a study detailing the link between bumblebee proboscis lengths and flowercorolla lengths (Pyke 1982). In this study, the long-proboscis bee species would preferentially feed on the long-corolla plants, the medium-proboscis bee species would feed on the medium-corolla plants, and so on. By evolving different proboscis lengths, several competing bee species are able to partition the available resources and coexist.

Conditional differentiation

The second form of niche differentiation is conditional differentiation, which occurs when two competing species differ in their abilities to use a resource based on varying environmental conditions. One species may be more competitive in one set of environmental conditions, but another species is more competitive in another set of conditions. Therefore, in a varying environment, each species is sometimes a better competitor and they can coexist. Differentiation based on environmental conditions is often difficult to separate from resource differentiation, and often conditional differentiation includes one or more types of resource partitioning.

Utilization of two resources (Tilman's R)*

The final type of niche differentiation is based on Tilman's (1977, 1990) notion that if two species are competing for the same exact resource then the ultimate winner will be the species which can deplete the resource the lowest, surviving on the lowest amount of the resource. This alone does not allow coexistence. However, if two species rely on two resources differentially, then coexistence is possible when each species can tolerate a lower amount of only one resource compared to its competitor. An example would be if grass species 1 is more limited by nutrient B than A and grass species 2 is more limited by nutrient A than B. Then, if species 1 uses more of nutrient B than A and species 2 uses more of nutrient A than B, species 1 can out-compete species 2 for nutrient A and species 2 can out-compete species 1 for nutrient B. Additionally, the starting point for each nutrient's availability must be roughly an equal distance from each species limits (i.e. no species can quickly lower one resource while the other is still in abundance). Coexistence is now possible because each species uses more of the nutrient that limits its own growth, but each still requires both nutrients to survive. This subtle form of niche differentiation is dependent on two conditions: 1) the habitat needs to be such that one species is more limited by one resource, and the other species is more limited by the other resource, and 2) each species must consume more of the resource that strongly limits its own growth.

Coexistence without niche differentiation: exceptions to the rule

Some competing species have been shown to coexist on the same resource with no observable evidence of niche differentiation and in "violation" of the competitive exclusion principle. One instance is in a group of hispine beetle species (Strong 1982). These beetle species, which eat the same food and occupy the same habitat, coexist without any evidence of segregation or exclusion. The beetles show no aggression either intra- or inter-specifically. Coexistence may be possible through a combination of non-limiting food and habitat resources and high rates of predation and parasitism, though this has not been demonstrated.

This example illustrates that the evidence for niche differentiation is by no means universal. Niche differentiation is also not the only means by which coexistence is possible between two competing species. However, niche differentiation is a critically important ecological idea which explains species coexistence, thus promoting the high biodiversity often seen in many of the world's biomes.

Research using mathematical modelling is indeed demonstrating that predation can indeed stabilize lumps of very similar species. Willow Warbler and Chiffchaff and other very similar warblers can serve as an example. The idea is that it is also a good strategy to be very similar to a successful species or have enough dissimilarity. Also trees in the rain forest can serve as an example of all high canopy species basically following the same strategy. Other examples of nearly identical species clusters occupying the same niche were water beetles, prairie birds and algae. The basic idea is that there can be clusters of very similar species all applying the same successful strategy and between them open spaces. Here the species cluster takes the place of a single species in the classical ecological models.

5

Reproductive Behaviour in Animal

In mammals, the major organs of the reproductive system include the external genitalia (penis and vulva) as well as a number of internal organs including the gamete producing gonads (testicles and ovaries). Diseases of the human reproductive system are very common and widespread, particularly communicable sexually transmitted diseases.

Most other vertebrate animals have generally similar reproductive systems consisting of gonads, ducts, and openings. However, there is a great diversity of physical adaptations as well as reproductive strategies in every group of vertebrates.

VERTEBRATES

Vertebrate animals all share key elements of their reproductive systems. They all have gamete-producing organs or gonads. In females, these gonads are then connected by oviducts to an opening to the outside of the body, typically the cloaca, but sometimes to a unique pore such as a vagina or intromittent organ.

Male

The male reproductive system is a series of organs located outside of the body and around the pelvic region of a male that contribute towards the reproduction process. The primary direct function of the male reproductive system is to provide the male sperm for fertilization of the ovum.

The major reproductive organs of the male can be grouped into three categories. The first category is sperm production and storage. Production takes place in the testes which are housed in the temperature regulating scrotum, immature sperm then travel to the epididymis for development and storage. The second category are the ejaculatory fluid producing glands which include the seminal vesicles, prostate, and the vas deferens. The final category are those used for copulation, and deposition of the spermatozoa (sperm) within the female, these include the penis, urethra, vas deferens, and Cowper's gland.

Major secondary sexual characteristics includes: larger, more muscular stature, deepened voice, facial and body hair, broad shoulders, and development of an adam's apple. An important sexual hormone of males is androgen, and particularly testosterone.

The testes release a hormone that controls the development of sperm. This hormone is also responsible for the development of physical characteristics in men such as facial hair and a deep voice.

Female

The human female reproductive system is a series of organs primarily located inside of the body and around the pelvic region of a female that contribute towards the reproductive process. The human female reproductive system contains three main parts: the vagina, which leads from the vulva, the vaginal opening, to the uterus; the uterus, which holds the developing fetus; and the ovaries, which produce the female's ova. The breasts are involved during the parenting stage of reproduction, but in most classifications they are not considered to be part of the female reproductive system.

The vagina meets the outside at the vulva, which also includes the labia, clitoris and urethra; during intercourse this area is lubricated by mucus secreted by the Bartholin's glands. The vagina is attached to the uterus through the cervix, while the uterus is attached to the ovaries via the fallopian tubes. Each ovary contains hundreds of egg cells or ova (singular ovum).

Approximately every 28 days, the pituitary gland releases a hormone that stimulates some of the ova to develop and grow. One ovum is released and it passes through the fallopian tube into the uterus. Hormones produced by the ovaries prepare the uterus to receive the ovum. The lining of the uterus, called the endometrium, and unfertilized ova are shed each cycle through the process of menstruation. If the ovum is fertilized by sperm, it attaches to the endometrium and the fetus develops.

Other Mammale

Didactic model of a mammal urogenital system. Most mammal reproductive systems are similar, however, there are some notable differences between the non-human mammals and humans. For instance, most male mammals have a penis which is stored internally until erect, and most have a penis bone or baculum. Additionally, males of most species do not remain continually sexually fertile as humans do. Like humans, most groups of mammals have descended testicles found within a scrotum, however, others have descended testicles that rest on the ventral body wall, and a few groups of mammals, such as elephants, have undescended testicles found deep within their body cavities near their kidneys.

The reproductive system of marsupials is unique in that the female has two vaginae, both of which open externally through one orifice but lead to different compartments within the uterus; males usually have a two-pronged penis which corresponds to the females' two vaginae. Marsupials typically develop their offspring in an external pouch containing teats to which their newborn young (joeys) attach themselves for post uterine development. Also, marsupials have a unique prepenial scrotum. The 15mm (5/8 in) long newborn joey instinctively crawls and wriggles the several inches (15 cm), while clinging to fur, on the way to its mother's pouch.

The uterus and vagina are unique to mammals with no homologue in birds, reptiles, amphibians, or fish. In place of the uterus the other vertebrate groups have an unmodified oviduct leading directly to a cloaca, which is a shared exit-hole for gametes, urine, and feces. Monotremes (i.e. platypusand echidnas), a group of egg-laying mammals, also lack a uterus and vagina, and in that respect have a reproductive system resembling that of a reptile.

Dogs

In domestic canines, sexual maturity (puberty) occurs between the ages of 6 to 12 months for both males and females, although this can be delayed until up to two years of age for some large breeds.

Horses

The mare's reproductive system is responsible for controlling gestation, birth, and lactation, as well as her estrous cycle and mating behavior. The stallion's reproductive system is responsible for his sexual behavior and secondary sex characteristics (such as a large crest).

Birds

Male and female birds have a cloaca, an opening through which eggs, sperm, and wastes pass. Intercourse is performed by pressing the lips of the cloacae together, which is sometimes known as intromittent organ which is known as a phallus that is analogous to the mammals' penis. The female lays amniotic eggs in which the young fetus continues to develop after it leaves the female's body. Unlike most vertebrates female birds typically have only one functional ovary and oviduct. As a group, birds, like mammals, are noted for their high level of parental care.

Reptiles

Reptiles are almost all sexually dimorphic, and exhibit internal fertilization through the cloaca. Some reptiles lay eggs while others are viviparous (animals that deliver live young). Reproductive organs are found within the cloaca of reptiles. Most male reptiles have copulatory organs, which are usually retracted or inverted and stored inside the body. In turtles and crocodilians, the male has a single median penis-like

Amphibans

Most amphibians exhibit external fertilization of eggs, typically within the water, though some amphibians such as caecilians have internal fertilization. All have paired, internal gonads, connected by ducts to the cloaca.

Fish

Fish exhibit a wide range of different reproductive strategies. Most fish however are oviparous and exhibit external fertilization. In this process, females use their cloaca to release large quantities of their gametes, called spawn into the water and one or more males release "milt", a white fluid containing many sperm over the unfertilized eggs. Other species of fish are oviparous and have internal fertilization aided by pelvic or anal fins that are modified into an intromittent organ analogous to the human penis. A small portion of fish species are either viviparous or ovoviviparous, and are collectively known as livebearers.

Fish gonads are typically pairs of either ovaries or testes. Most fish are sexually dimorphic but some species are hermaphroditic or unisexual.

INVERTEBRATES

Invertebrates have an extremely diverse array of reproductive systems, the only commonality may be that they all lay eggs. Also, aside from cephalopods, and arthropods, nearly all other invertebrates are hermaphroditic and exhibit external fertilization.

Cephalopods

All cephalopods are sexually dimorphic and reproduce by laying eggs. Most cephalopods have semi-internal fertilization, in which the male places his gametes inside the female's mantle cavity or pallial cavity to fertilize the ova found in the female's single ovary. Likewise, male cephalopods have only a single testicle. In the female of most cephalopods the nidamental glands aid in development of the egg.

The "penis" in most unshelled male cephalopods (Coleoidea) is a long and muscular end of the gonoduct used to transfer spermatophores to a modified arm called a hectocotylus. That in turn is used to transfer the spermatophores to the female. In species where the hectocotylus is missing, the "penis" is long and able to extend beyond the mantle cavity and transfer the spermatophores directly to the female.

Insects

Most insects reproduce oviparously, i.e. by laying eggs. The eggs are produced by the female in a pair of ovaries. Sperm, produced by the male in one testis or more commonly two, is transmitted to the female during mating by means of external genitalia. The sperm is stored within the female in one or more spermathecae. At the time of fertilization, the eggs travel along oviducts to be fertilized by the sperm and are then expelled from the body ("laid"), in most cases via an ovipositor.

Arachnids

Arachnids may have one or two gonads, which are located in the abdomen. The genital opening is usually located on the underside of the second abdominal segment. In most species, the male transfers sperm to the female in a package, or spermatophore. Complex courtship rituals have evolved in many arachnids to ensure the safe delivery of the sperm to the female.

Arachnids usually lay yolky eggs, which hatch into immatures that resemble adults. Scorpions, however, are either ovoviviparous or viviparous, depending on species, and bear live young.

Terrestrial Animals

Haeckel (1866) first defined Ecology as 'the science of relations between organisms and their environment' (Bramwell 1989). The study of ecology has developed over the years from an initial descriptive field of study in the 19th century to a more quantitative, experimental and analytical discipline in the 21st century (Mayorga et al. 2002). The definition has also developed. Krebs (1972) defined it as 'Ecology is the scientific study of the interactions that determine the distribution and abundance of organisms'. It is clear however that the interactions to which he referred are the very factors which create the environment and so a more complete definition which marries the definitions given by Haeckel (1866) and Krebs (1972) is suggested as follows by Begon et al. (2006): 'Ecology is the scientific study of the distribution and abundance of organisms and the interactions that determine distribution and abundance.'

It is necessary to take a historical aspect to understand the present distribution of plants and animals within the United Kingdom. The initial event is generally seen to be the movement of land masses followed by several periods of glaciation. Pleistocene glaciation is largely the event determining the present day patterns in distribution of Flora and Fauna in the UK (Begon et al 2006). Most of the plants and animals were pushed further south due to the icy conditions. Recolonisation of plants and animals is dependent on distance from other populations and the ability of some animals to disperse and reproduce, grow and survive, once they arrive in a suitable environment. There are two notable contrasts with continental Europe, 'Britain has a relatively impoverished mammal fauna as several species, such as the garden dormouse Eliomys quercinus and the beech marten Murtesfuina, now found on the western edge of the continental mainland, failed to recolonise Britain after the last ice age' (Mitchell-Jones 1996). Reptiles and amphibians species are also more diverse on the continent than in the UK. There are only 6 native species in the United Kingdom, by comparison with a single species in Ireland which is further from Europe, Europe itself boasts over 87 reptilian species (Silva et al. 2009).

A niche is a summary of the organism's tolerances and requirements, how they interact to define the conditions and resources needed by an individual or species in order to practice its way of life (Begon et al. 2006), and the time it occurs there (Mackenzie et al. 1998). The niche of an animal is generally larger than that it actually inhabits, This is the fundamental niche which is characterised by conditions (temperature, relative humidity amongst others), within the tolerable limits of a species provided that there are enough resources available i.e. food, accommodation and that it is not limited by interactions with other organisms such as predation and competition and that it is not prevented from getting to the suitable area (dispersal). Hutchinson (1957) identified the Fundamental niche from the realised niche with the latter being "the more limited spectrum of conditions and resources that allow an animal to persist even in the presence of competition and predation.

Conditions are defined as variable environmental factors which organisms respond to. They are non depletable, the organism cannot use them up (Mackenzie et al. 1998) These are portrayed as one dimensional but clearly the effect of one condition may be altered by another i.e. Cool winds can high temperatures more bearable, thus making the effects often two or indeed multidimensional.

Temperature is a condition which affects the rate of development in organisms, in fact many cold blooded animals incubation and development is given in degree days rather than in actual time. An example of how temperature affects not one but ultimately two species in their realised niche and distribution is given by Randall (1982). The rush moth (Coleophora alticolella) lays its eggs on the flowers of the rush Junctus squarrosus and the caterpillar uses the ripe seeds as its food resource. The moths and the larvae are little affected by low temperatures so there is no reason why they can't extend their niche further up in altitude however at the lower temperatures above 600m the seed of the rush fails to ripen and so there is no food resource for the caterpillar. The temperature related limit of tolerance has been reached for the rush, which in turn limits the niche of the caterpillar giving us the realised niche. Relative humidity plays a very important part in the distribution of animals but it is impractical to discuss it in isolation due to the large influence that temperature has on it. Relative humidity affects the rate at which we lose water. Increasing temperature increases humidity, the physiology of animals is important with this limiting condition. 'The animals most affected by humidity are those 'terrestrial' animals that are actually, in terms of the way they control their water balance' 'aquatic'. (Begon et al. 2006).

The pH of soil and water can have a strong influence on plant and animal communities. Plant roots tend to be damaged in soils below pH3 and above pH 9 due to the pathogenic effect of toxic levels of H+ or OH- ions. Soil pH also has an effect on the uptake of nutrients and the concentration of toxins, tolerance levels vary for pH but only a minority of plants can grow at pH less than 4.5 (Begon et al. 2006). Kidd and Proctor (2001) observed separate adaptations in the various populations of the grass Holcus lanatus L. (Yorkshireâ fog) and the tree Betula pendula Roth (Silver Birch) in very acid soils at some Scottish sites to H+ or Al3+ toxicity. These were closely related to the edaphic characteristics of the original site from which they were collected. 'These adaptations support the view that this is an important factor in very acid soils' (Kidd and Proctor 2001).

Increased salinity in the soil water offers osmotic resistance to water uptake. Wind plays a major role in plant dispersal. Small light seeds have developed special adaptations which facilitates there dispersal by the winds. These adaptations are to enable the seeds to remain airborne longer so they can be carried greater distances. These adaptations increase the area to catch the wind.

Sycamore

The fruit (which contains the seed) delays its fall so that the wind may carry it a far distance away. (Photo:http://www.countrysideinfo.co.uk/seed_dispersl/wind.htm)Wind also has a negative effect on the distribution of organisms in that it has been implicated in the problem of soil erosion of arable lands and of sand dunes, thereby reducing their capacity to sustain plant communities.

While Man has a positive impact in the management of resources and in the study of ecosystems, man is also the perpetrator of the majority of effects leading to environmental pollution. These include the clearing of land for operations totally wiping out local habitats e.g. Corn Crake (RSPB 2010), the leaching of metals into the environment and the dumping of copper, zinc and lead around mines, elevating their presence to lethal levels (Begon et al..2006). Power plants and other factories may emit sulphur dioxide and nitrogen oxide which facilitate the problems with acid rain. Researchers now know that acid rain causes slower growth, injury, or death of forests. It is practical to assume that if it has this effect in forests it may also retard the growth of other plants which affects a food resource (National Geographic 2010).

When pollution occurs, organisms often find away to combat the stress and overcome the effects this is evidenced in the well known example of the peppered moth. However while the development is a reaction to industrial pollution there are many other factors at play, such as genetic variability. With the increasing industrialisation in Britain, the peppered moth survived by developing a darker coloured form which was better camoflaged from predators when it landed on the soot darkened trees after the lichens had died off (Majerus and Stevens 2006).

Environmental pollution leads us then to consider the changes we are making that ultimately influence these conditions termed as global warming. This is occurring at a rate that will not allow flora or fauna to adapt and thus could change living conditions so to be intolerable to many species. Berry et al. (2003) have undertaken a study which shows the vulnerability of terrestrial habitats and species distribution in Britain to climate change which is essentially the temperature increase of 0.6 C over the past century. They contend that with such changes that it is not safe to assume that a species historical range of distribution will remain suitable.

Resources are all things used by an organism, some are infinite, others may involve competition, but in all cases resources that are used up by the organism are not available to others. Resources can be places to live and many other specific items used by organisms but for the present we will discuss

Energy: Solar radiation

CO2

Water

Oxygen

Plants are the primary producers and are the foundation of the food chain/web for all animals. Plants need energy; they can only get this from solar radiation. Radiation must be captured and chemically fixed by plants or it is lost forever as energy (Begon et al.). When radiation is used by one plant it is no longer available to others. It is not recycled like nutrients, carbon and minerals. Radiation is important to most plants along with CO2 and water for photosynthesis. It is difficult to discuss radiation separate to Water and CO2 as one often influences the other.

Water is a vital resource essential for all organisms. It is essential for photosynthesis in plants and for transportation of nutrients throughout the plant. It is critical for animals for respiration and again for transportation and for many of the metabolic processes. The degree of morphological and physiological specialization that has occurred in plants to adapt to radiation, intensity and water availability is vast and too extensive for present discussion.

CO2 is vital for photosynthesis by plants and is drawn from the air it is released by animals after respiration and by plants at night. Levels in a terrestrial environment vary greatly (Begon et al. 2006) and so organisms have developed different ways to enhance its utilization.

Carbon dioxide is the single resource which is increasing globally, due to man's clearing of forests and burning of fossil fuels.

Oxygen is critical for all plants and animals. Only a small number of bacteria can survive without it.

When resources are limited or affected by other factors they produce stress in organisms. This generally leads to a reduction in abundance at the extremes of tolerance. Tolerance can be explained as the peak of endurance of species to an extreme bearable condition. When this situation avails, the species limit of tolerance in which the organism can survive and grow in that environment has been stretched. This can also be regarded as ecological stress zone. 'As conditions approach the upper and lower extremes of environmental conditions which individuals can survive the species of tolerance and no growth occurs'(McKenzie et al. 1998).

There are many other factors which influence the fundamental niche, and the distribution of organisms in the UK, these are interactions with other organisms such as predation and competition. Where a natural predator is removed the prey can thrive so well that it may be in danger of destroying its own environment. Such is the case with the Stag in the highlands. With no natural predator its population is at its highest level in over a thousand years (Postnote 2009). If an organism is introduced that is similar to a pre-existing organism it may out compete it and cause patchiness in its distribution, this is evident in the competition between the introduced grey squirrel and the native red squirrel (Okubo et al. 1989). As all organisms are a potential food source for others, many animals defend themselves through physical, chemical, camouflage,

The main ecological factors effecting the distribution of organisms in the UK are as outlined above. The temperate nature of the United Kingdom appears not to be extreme; however it is pertinent to note that the relativeness of the ecological interraction of the organisms on above factors make it more complex and to the effects of the distribution of terrestial plants and animals in United Kingdom. A single leaf is important to an insect and may possibly sustain the larva for its whole existence but it is of no effect to a cow. We have seen from the earlier example of the rush moth how distribution of one species can be a determining limit tolerance factor level of another species.

Growth and Development in Animal

Cloning is the creation of an organism that is an exact genetic copy of another. This means that every single bit of DNA is the same between the two organisms. Cloning is at present illegal because of the devastating impacts it may have on humanity, cloning became popular when scientist Ian Wilmut first cloned Dolly by a process termed somatic cell nuclear transfer. Dolly was the only success out of 277 experiments and Cumulina, the first cloned mouse after Dolly, was among 15 live born mice from 942 tries; so this proves that cloning is yet not quite understandable for we have little knowledge about it. But even so humans won't stop until it is a safe success. Cloning is bound to bring great benefits to human kind but it will also bring fort great chaos from ethical views and religion. The idea that a clone is an exact duplicate of another individual is not reliable, and so if the intent of cloning is to create such a copy, it simply will not work. To be clear, the tips of chromosomes, called telomeres, shorten with each cell division. A clone's telomeres are as short as those from the donor nucleus, which means that they are "older" even at the start of the clone's existence. DNA in the donor nucleus has also had time to mutate, that is to say, it has had time to undergo modification from its original sequence, thus distinguishing it genetically from other cells of the donor. A mutation that would have a negligible or delayed effect in one cell of a many-celled organism, such as a cancer-causing mutation, might be devastating if an entire organism develops under the direction of that nucleus. Finally, the clone's mitochondria, the cell organelles that house the reactions of metabolism and contain some genes, are those of the recipient cell, not the donor, because they reside in the cytoplasm of the egg. Mitochondrial genes, therefore, are different in the clone than they are in the nucleus donor. The consequences of nuclear and mitochondrial genes from different individuals present in the same cell are not known, but there may be incompatibilities. Perhaps the most compelling reason why a clone is not really a duplicate is that the environment affects gene expression. Cloned calves have different colour patterns, because when the animals were embryos, the cells that were destined to produce pigment moved in different ways in each calf. For humans, consider identical twins. Nutrition, stress, exposure to infectious diseases, and other environmental factors greatly influence our characteristics. For these reasons, cloning a deceased child, the application that most would-be cloners give for pursuing the technology, would likely lead to disappointment.

Actual Process

In its most basic procedure, cloning encompasses three footsteps. In the first step, scientists obtain cells from an individual whose characteristics they want to copy for example they want to copy an ordinary dog's colour, consider that I have a dog pot-licker but he has his two front legs white with black spots, like a dalmata. It's this characteristic that they want to copy so they would take his cells for the process. They then place these cells, which are called donor cells, into a liquid culture. This culture contains nutrients that stop the cells from dividing. In the second footstep, an unfertilized egg is taken from a female. This unfertilized eggs nucleus is then removed, leaving an empty egg cell. The donor cell is then taken form the culture and is placed into the empty egg. This progression creates an embryo that is an exact copy of the donor, which would be the spotted leg dog and not the mother. In the final step, the embryo is put into the uterus of a female of the same species and the offspring arrives into the world by means of the natural birth process.

Plants

From a long time ago, people have been taking roots or stems of plants in order to make genetically identical copies. Usually, this is done by choosing the best plant, of course (for example, the most decorative or unusual), cutting a root or branch from it, and placing that cutting in water or soil. The cells will then divide and double in size every six weeks until the cutting develops roots. At this point, it is ready to be planted. It will then grow into an exact copy of the parent plant. This is normally done with flowers and vegetables and fruits.

Animals

Dolly, a Finn Dorset sheep, the first mammal cloned from an adult cell, was produced in Scotland at the Roslin Institute in Edinburgh, this happened in July 1997. Ian Wilmut and colleagues informed the birth of a lamb resulting from the transfer of a nucleus obtained from an adult ewe udder cell into an enucleated oocyte that was then implanted into a surrogate mother. Dolly was cloned through a process called nuclear transfer. The process that produced Dolly differs from ordinary reproduction in two major ways. First, body (or somatic) cells from an adult ewe's udder (this is the donor) were placed in a culture dish and it was allowed to grow. The nutrients were then removed from the culture, which stopped the cells' growth. One of these nongrowing cells was then fused by electric jolts with another ewe's oocyte from which the nucleus had been previously removed. This procedure is known as 'somatic cell nuclear transfer'. Within a day the fused cells began to divide in the culture dish. After several divisions, the early embryo was transferred to the uterus of a surrogate mother and allowed to develop. Second, unlike the sperm and the egg, each of which contributes half the number of chromosomes at fertilization, each body cell contains twice the number of chromosomes in each germ cell. So fusion of a sperm and an egg forms an individual whose full genetic composition is unique to that individual. On the other hand, the embryo cloned from somatic cell nuclear transfer begins development with the diploid (double) number of chromosomes, all derived from one somatic cell (adult udder) of a single individual. This embryo has the same nuclear genetic composition as the donor of the somatic cell. In the end, three sheep contributed to the production of a single lamb clone: a Finn Dorset sheep donated her udder cells for culture; a Scottish Blackface sheep donated the enucleated oocyte (with its nucleus removed, thus losing its own genetic identity in the process); and a Scottish Blackface sheep became the surrogate mother, carrying the embryo to birth.

There are 4 type of cloning: recombinant DNA technology or DNA cloning, reproductive cloning, therapeutic cloning and replacement cloning.

Therapeutic cloning, also called "embryo cloning," is the production of human embryos for use in research. The goal of this process is not to create cloned human beings, but rather to harvest stem cells to be used to study human development and to treat disease.

Reproductive cloning is a technology used to generate an animal that has the same nuclear DNA as another currently or previously existing animal.

The terms "recombinant DNA technology," "DNA cloning," "molecular cloning," and "gene cloning" all refer to the same process: the transfer of a DNA fragment of interest from one organism to a self-replicating genetic element such as a bacterial plasmid. The DNA of interest can then be propagated in a foreign host cell.

CONCEPTS & SIGNIFICANCE

Cloning is seen as the way to keep our species survival going, a way to better human life, a way to evolve, a way to relief, it is seen as the way out of many of our troubles. But cloning is also seen as the interference of nature; the meaning of cloning for people that see it as nature's violation is great chaos and corruption of mankind by creating deformed beings which would then be living with the deformities. The concept of rejecting the idea of cloning is often based on emotions which stop humans from improving. Other concepts of cloning are that these clones would be treated as chemical ingredients this is considering the harvesting of organs for transplants and of course killing the clones. Clones might also be regarded as subjects rather than lives to their originals. Genetic cloning is very important. I don't understand why people would not agree with it, maybe because they don't know what it is. Genetic cloning is not cloning whole humans or any other animals, yet it most certainly could. Genetic cloning allows the experimentation and study of genetic disorders. By performing these experiments we learn about the different processes that occur within the cell thus the more we learn about these processes the more we can do to prevent disorders that occur due to malfunctions in these processes. Some people think that cloning is a "what the hell idea" I don't believe so.

Benesits to Humanity

Cloning could be used to produce vital organs for transplant. The only way to do this, however, would be to clone the entire individual, including its organs. This practice would raise ethical questions and would require a long time to grow the organism. It would take a long time for a donor's organs to be mature enough to be removed from the donor and used for transplant. In addition, scientists are unsure whether transplanted organs from cloning would be accepted or rejected by the recipient individual. Another benefit to humanity would be the salvation of endangered species. At that present time, the success was not likely to happen because it took 277 tries to clone Dolly the sheep.

But if the success rate of cloning increases, it could be a way to increase the population of endangered species or animals that are difficult to breed for example the panda. Extinct species could also be revived using the cloning method although this would be more difficult. Cloning extinct animals has two problems. First, donor cells must be taken from living organisms; unless an extinct animal is found frozen (the woolly mammoth recently discovered in the Arctic), it would be impossible to find living cells. Because the fossil bones of dinosaurs contain no living cells, a dinosaur cannot be cloned. Second, current cloning technology requires a surrogate mother and an egg cell from a living female of the same species, which is something that we don't have yet. Females normally cannot give birth to an animal from a different species.

It is unlikely, for instance, that a female elephant could donate an egg cell and give birth to a woolly mammoth. Other benefits for us would be the obtaining of desired traits through cloning than through conventional breeding. For example, cloning could benefit crop engineering by creating foods that are more nutritious, disease free, and plentiful. Cloning could also help in the prevention and cure of diseases. In example, the same laboratory where Dolly was created is now working to create eggs that contain anticancer proteins to prevent various forms of cancer (such as fast growing forms of skin cancer). Dolly herself was cloned to produce a sheep whose milk had more proteins that are believed to help treat diseases such as emphysema, haemophilia, and cystic fibrosis. An interesting benefit is that aged cell nuclei can be rejuvenated. Studies using cell culture have shown that body cells grow and divide normally in culture for a while, but eventually stop dividing, become senescent, and die. An exception was seen in aged frog red blood cell nuclei (human red blood cells lack nuclei).

After their transfer into enucleated oocytes, frog red blood cell nuclei were rejuvenated. They carried out the formation of tadpoles that survived almost a third of the way to metamorphosis. The oocyte cytoplasm contains an abundance of chemicals that promote DNA synthesis and cell division after normal fertilization. It's believed that these substances also rejuvenate aged cell nuclei and turn non-cycling frog red blood cells into active ones. If they could isolate these substances, they might be able to alleviate or reverse senescence. Cloning will allow us to improve human life. It will allow more individuals to live more rewarding lives. And, it will actually have the potential to increase the quality of life of those who do not rank as high on the principal valued dimensions as the clones themselves. In addition, it will give us earlier knowledge of medical conditions of clones, so that these conditions can be addressed before they become life threatening. Longevity will also be increased by other means (chiefly gene therapy and anti-aging drugs).

Thus, if the clones of a certain person have a normal life expectancy of 90 years, by the time they reach this life expectancy, in all probability, medical technology will have advanced to the point where another 30 or more years can be added to this. Hence, a clone's foreknowledge that their proto died at a certain age from natural causes will not be an automatic death sentence for them. Clones can expect to add years to their lives through environmental influences and medical advances, so the age of their natural death will remain uncertain. There are many additional benefits from cloning to be considered. Cloning has great implications for the human species to the extent that candidates can be selected which are largely free from genetic defects. Correspondingly, the non-cloned offspring of clones will be more likely to be free from genetic defects themselves, thereby improving the quality of the human gene pool.

The fears associated with establishing a qualitative distance between clones and the rest of humanity have already been discussed at some length. But summarily, the tendency of cloning will be to push the entire species in the direction of greater functionality and survivability. Inevitably, clones will marry and propagate with non-clones that are not their equals. In effect, some clones will marry to become weaker. Correspondingly, some non-clones will marry to become stronger, resulting in a genetic improvement and uplifting of the entire human race. In short, cloning will allow us to accelerate our evolution beyond what would normally occur merely by means of the choices people make in the mate selection process. Pet lovers would also have their pets for as long as they want by cloning and unfertile couples would have babies too.

Ethical Issues

The arrival of Dolly sparked widespread rumours about a human child being created using somatic cell nuclear transfer. Much of the superficial fear that met this announcement centred on the misunderstanding that a child or many children could be produced who would be identical to an existing person. This fear was based on the idea of "genetic determinism" genes alone determine all aspects of an individual and reflects the belief that a person's genes bear a simple relationship to the physical and psychological traits that make him or her. Genes play an essential role in the formation of physical and behavioural characteristics; although each individual is, in fact, the result of an interaction between his or her genes and the environment within which he or she develops. Many of the concerns about cloning have focused on issues related to the idea of "playing God," this is considered an interference with the natural order of life. Also there are those who believe that the embryo has the moral status of a person from the moment of conception, research or any other activity that would destroy it is wrong. Cloning ethics is structured in seven ways

The potential impact of cloning on individuals; its potential to create a genetic underclass; Clones would be considered better than others or even us humans.

The potential impact of cloning on the social structure and the division of labor. Some clones would be used for labour which would be wrong.

The implications of cloning for the composition of the gene pool and the future of human race

Who decides who is eligible to be cloned?

How will the decision to select and approve candidates for cloning be made, and what criteria will be used?

What will be the quality of life of the clones?

What are the implications of cloning for the survivability of the species; how will it enhance survivability?

Cloning has opened many doors and has lit many roads that could lead to significant medical advancements but, as with all new technologies, it will be accompanied by ethical and social standoffs. Today's successes will shape the road to improving efficiencies and help add to the basic understanding of our cells. Even Dolly's creator, Ian Wilmut, is focusing less on sheep and more on understanding the mechanism of reprogramming our genetic material. In conclusion, cloning is unsafe at this time because the complete list of defects cannot be accounted for. The sacrifice of countless children to death and physical abnormalities, as well as mental illness, cannot become real out for the sake of science. Experts say that cloning humans in the near future is not safe because: "It seems that a little knowledge is a dangerous thing and the authors (other scientists) have allowed themselves to over-interpret their findings," said Ian Wilmut. This is a science that cannot be rushed into, because the consequence is paid with a lifetime of grief for a defective child.

ADVANTAGES AND DISADVANTAGES SEXUAL REPRODUCTION IN ANIMALS

Reproduction is the production of new individuals by an existing member or members of the same species. It is a fundamental characteristic of living things; the ability to self replicate in this way sets the living world apart from the non-living. Sexual reproduction is the exclusive form of reproduction in all higher invertebrates and in all vertebrates, whereas in few parthenogenesis is also possible. Also it occurs in many multicellular and in all unicellular organisms such as plants and animals. In sexual reproduction one sex cell (egg) fuses with another sex cell (sperm) producing a new cell called zygote. Fusion of gametes or sex cells is called fertilization. In the process of gamete formation nuclear division known as meiosis halves the normal chromosome number. That is gametes are haploid and fertilization restores the diploid number of chromosomes. There are many advantages and disadvantages in sexual reproduction; these will be discussed further in this essay.

One of the most important advantages of sexual reproduction is that it results in genetic variation among offsprings. This is a result of meiosis, which is the halving of the number of chromosomes and fertilization, whereby the male and female sex cells fuse and the restoration of the original number of chromosomes. Among them meiosis is the major source of genetic variation. This genotypic variation is caused due to the two events; independent assortment and crossing over that take place during meiosis. Independent assortment occurs because the way the bivalents line up at the equator of the spindle during meiosis Ðâ is entirely random. Therefore, the number of possible combinations of chromosomes that can be formed by random orientation in meiosis is 2n (n= number of pairs of chromosomes). There is also crossing over of segments of individual maternal and paternal homologous chromosomes (Hered, 1992).These events results in new combinations of genes on the chromosomes of the haploid cells produced. According to Hered (1992) these variations found in offspring of sexual reproduction allow some individuals to be better suited for survival and provide a mechanism for selective adaptation to occur.

Genetic recombination is another advantage of sexual reproduction. According to Antonovics & Kelly (1988) genetic recombination helps to separate beneficial mutations from harmful mutations. An organism can be subjected to many changes due to mutation. Most of the non-neutral mutations are known to be deleterious and this means that they will cause a decrease in organisms overall fitness. Mutations with deleterious effects can be removed from the population by natural selection (Antonovics & Kelly, 1988).It is believed that natural selection acts upon the entire genetic makeup of an organism without sexual reproduction, but with sexual reproduction natural selection acts upon individual genetic traits.

Sexual reproduction also helps in the spread of advantageous traits. In sexual reproduction novel genotype are produced. As mentioned earlier since this type of reproduction is due to the fusion of two gametes with different combination of genes, populations are capable of combining advantageous genes than in asexual (Crow, 2005). Sexual reproduction increases resistance to parasites, this is also known as the Red Queen hypothesis. The Red Queen refers to the frequency-dependent relationship between host and parasites. It shows those sexual hosts are continually adapting in order to stay ahead of their parasites. Thereby, this clearly illustrates that sexual reproduction helps to increase parasitic resistance (Kang, 2008,).

Disadvantages of Sexual Reproduction

However, sexual reproduction is not the perfect reproductive method if it examined in detail there are many disadvantages which are considerable. As mentioned by Graham (1982) one f the draw backs of sexual reproduction is that the cost of reproducing offspring by means of sexual reproduction is almost twice the cost of reproducing asexually. That is while sexual lineage is completing one division cycle, asexual lineage could finish several division cycles. This is because meiosis takes considerably longer time than two mitotic divisions. The processes of syngamyi and karyogamy (combining of gametic nuclei) also require a longer period. As explained by Crow (2005) Chromosomes recombination and independent assortment are the major features of meiosis and as a result of these features the allelic combinations can be disrupted within and between chromosomes. In other words, in sexual reproduction they are incapable of producing clones as offspring. Therefore, because of this process it is likely that high fitness genotypes will be destroyed and the mean fitness of the offspring will be reduced.

Another disadvantage is the additional expenditure of time and energy needed to attract or find mates. Also during that time spent an individual may encounter predators (Campbell & Reece, 2008). Furthermore, as pointed out Kondrashov (2001) ââ Å"cost of meiosis or 50% cost, shows that for example if there is, a population of 100 asexual and 100 sexual individuals in the same environment, while only 50% of the sexual population can produce offsprings, all the asexual population will be able to produce offsprings. This is known as 2:1 demography disadvantage.

Unlike in asexual reproduction there is another huge risk of exposure to sexually transmitted diseases during sexual reproduction in animals as well. It is clearly said that the time during mating is the ideal time for the transmittance of parasites and pathogens form one individual to another. The most common sexually transmitted disease among animals today is brucellosis (undulant fever).This is common among animals such as dogs, goat, rats and domestic livestocks. (Oriel & Hayward, 1974).

In conclusion, sexual reproduction brings about many advantages such as genetic variation, recombination function to eliminate mutations, also increase resistance to parasites and producing organisms different to each other. Though sexual reproduction results in disadvantages such as energy wasted during searching for a mate, time consumed during long cell cycles. Overall when it is compared to asexual reproduction, sexual reproduction results in far more advantages in the long run.

Cloning in Animals

What is cloning? Cloning is a process carried out in a laboratory by which a genetically identical organism can be made through non-sexual means. How it started? In February 1997, when embryologist Ian Wilmut and his colleagues at Roslin Institute in Scotland were able to clone a lamb, named Dolly. Who are cloned? Both plants and animals can be cloned. Plant Cloning is an ancient form of producing desired outcomes in plant species. Animal Cloning is the process by which an entire organism is reproduced from a single cell taken from the parent organism and in a genetically identical manner. This simply means the cloned animal is an exact duplicate in every way of its parent; it has the same and exact DNA. What does it involve? For different purposes there exist different extents of cloning such a molecular cloning; the process of making multiple molecules, Cellular cloning, which can be further divided into unicellular cloning; in which you derive a population of cells from a single cell, and Cloning in stem cell research; used to create embryos for research or therapeutic purposes, and the most complex type of cloning is organism cloning; it refers to the procedure of creating a new multicellular organism, genetically identical to another. Is cloning advantageous or disadvantageous? Cloning has both advantages and disadvantages to human beings. Advantages may include the creation of human beings that can be used to study human development and to potentially treat diseases. Disadvantages may include the lack of diversity of organisms and the hindrance of evolution. Why clone, and why not to clone? The ethics of cloning has become a great issue. Scientifically speaking, cloning can be the solution to problems associated with organ transplants that can benefit human health, and can also help in great yield production of healthy crops. But, ethically speaking the bible states that a new individual should be born from a husband (which provides the sperm) and a wife (which provides the egg), therefore by cloning one will blunder the world of God for it involves asexual reproduction of organisms. Cloning has indeed been the subject of scientific experiments for years but it was first known to be successful in 1997 when, after about 277 eggs used, the first cloned mammal 'Dolly' was born. In the United States, the human consumption of meat and other products from cloned animals was approved by the FDA on December 28, 2006, with no special labeling required.

Concept and Significance

Cloning can have slightly different meanings depending of the specific field one is studying. In general Biology cloning refers to the process of production of similar populations of genetically identical individuals that occur in nature when organisms such as bacteria, insects or plants reproduce asexually. In Biotechnology cloning refers to the process used to create copies of DNA fragments, cells, or organisms. Cloning has two main branches. They are Plant Cloning and Animal Cloning. Plant Cloning is an ancient form of producing desired outcomes in plant species. For farmers plant cloning is used to duplicate the success of good plants, and rich crops with very high yields. Plant cloning is easier than using seeds to generate new plants because the duplication of desired genes is rapid and the grower knows the type and quality of plant he would be growing since the new plant can be created from just a twig. Animal Cloning is the process by which an entire organism is reproduced from a single cell taken from the parent organism. This new organism is genetically identical; the exact duplicate in every way including DNA, to the parent organism. Animal cloning can be done both for reproductive and non-reproductive or therapeutic purposes. In the second case, cloning is done to produce stem cells or other such cells that can be used for therapeutic purposes, for example, for healing or recreating damaged organs. Cloning can happen both naturally and in the lab. Natural forms of cloning can include asexual reproduction in certain organisms such as bacteria and also in the development of twins from a single fertilized egg. Cloning can be done in the labs; an example would be the process of nuclear transfer of embryonic cells which leads to the production of cloned mammals. Cloning is divided into three major compartments, they are; molecular cloning, cellular cloning, and organism cloning. Molecular Cloning is the process of making multiple molecules. It is used to amplify DNA fragments containing whole genes, but can be used to amplify any DNA sequence such as promoters, non-coding sequence and randomly fragmented DNA. It is used for biological experiments and practical applications ranging from genetic fingerprinting to large scale protein production. Cellular Cloning has two subgroups; unicellular cloning and cloning in stem cell research. Unicellular cloning occurs in organisms such as bacteria and yeast; these processes are simple and only require inoculation of the appropriate medium for them to derive a population of cells from a single cell. For stem cell research, Somatic-cell nuclear transfer is

used to create embryos for the research and therapeutic purposes. This is called research cloning and its goal is to harvest stem cells that can be used to study human development and to potentially treat diseases such as Diabetes and Alzheimer's. Organism Cloning refers to the procedure of asexually creating a new multicellular organism, genetically identical to another. Successful animal cloning, in particular cloning from an adult animal, one knows exactly how the clone is going to turn out. This becomes especially useful when the whole intention behind cloning is to save a certain endangered species from becoming totally extinct.

ACTUAL PROCESS

Animals: Molecular Cloning which is the cloning of any DNA fragment can be divided into the following different steps:

Fragmentation - breaking apart a strand of DNA

Ligation - gluing together pieces of DNA in a desired sequence

Transfection - inserting the newly formed pieces of DNA into cells

Screening/Selection - selecting out the cells that were successfully transfected with the new DNA.

Unicellular Cloning of organisms is performed using a culture technique which involves the use of cloning rings. In this technique:

A single-cell suspension of cells that have been exposed to a mutagenic agent or drug used to drive selection is plated at a high dilution to create isolated colonies; each arising from a single and potentially cloned distinct cell.

At an early growth stage when colonies consist of only a few of cells, sterile polystyrene rings (cloning rings), which have been dipped in grease are placed over an individual colony and a small amount of trypsin is added.

Cloned cells are collected from inside the ring and transferred to a new vessel for further growth.

Somatic-Cell Nuclear Transfer is used to create embryos for research and therapeutic purposes. The SCNT process to create embryos for stem cell research is:

First, the collection of cells from the organism that will be cloned, the cells can be used immediately or can be stored in the lab for later use.

The maternal DNA from an oocyte is removed at metaphase II. The nucleus can then be inserted into an egg cytoplasm. This creates a one-cell embryo. The grouped somatic cell and egg cytoplasm are then introduced to an electrical current. The sexual energy allows the cloned embryo to begin development. The successfully developed embryos are then placed in surrogate recipients. Organism Cloning the procedure of creating a new multicellular organism, genetically identical to another is achieved by: The transfer of a nucleus from a donor adult cell (somatic cell) to an egg that has no nucleus. When the egg begins to divide normally it is transferred into the uterus of the surrogate mother.

An example of the first mammal to be successfully cloned from an adult cell is Dolly, a Finn-Dorset ewe. She was cloned at the Roslin Institute in Scotland. Dolly was formed by taking a cell from the udder of her biological mother. Her embryo was created by taking the cell and inserting it into a sheep ovum. The embryo was then placed inside a female sheep that went through normal pregnancy. She lived for 6 years, from 1996-2003. There were early claims that Dolly the Sheep had pathologies resembling accelerated aging, but other researchers, including Ian Wilmut who led the team that successfully cloned Dolly, argue that Dolly's early death due to respiratory infection was unrelated to deficiencies with the cloning process.

In amphibians the ultimate test of whether the nucleus of a differentiated cell has undergone any irreversible functional restriction is to have that nucleus generate every other type of differentiated cell in the body. If each cell's nucleus is identical to the zygote nucleus, then each cell's nucleus should be totipotent (capable of directing the entire development of the organism) when transplanted into an activated enucleated egg. Before such an experiment could be done, three techniques for transplanting nuclei into eggs had to be perfected: a method for enucleating host eggs without destroying them; a method for isolating intact donor nuclei; and a method for transferring such nuclei into the host egg without damaging either the nucleus or the oocyte.

Disadvantages of Cloning

In a large percentage of cases, the cloning process fails in the course of pregnancy or some sort of birth defects occur, for example, as in a recent case, a calf born with two faces. Sometimes the defects manifest themselves later and kill the clone. This long lasting process before the success of cloning can lead to the extinction of little remains of DNA from extinct species if it is used with no successive results.

Changes in genomes may not only result in changes in appearance, but in psychological and personality changes as well.

In plants; cloning limits diversity which makes the plants more susceptible to diseases and pests.

Possible potential harms and disadvantages are: the possibility of compromising individualities, the loss of genetic variation, technology is not well developed; it has low fertility rate; in cloning Dolly, 277 eggs were used, 30 started to divide, nine induced pregnancy, and only one survived.

Ethical Issues

Although most scientists consider the process of animal cloning as a major breakthrough and see many beneficial possibilities in it, many people are uncomfortable with the idea, they say it is 'against nature' and it is ethically damning, particularly in the instance of cloning human beings.

One of the main goals of the government is to protect human life. Some people want the government to regulate cloning and not allow it. Producing clones for research or to use their parts is unethical. It would be against the code of ethics of a doctor to harm a clone (i.e., use it for an organ transplant). The clone would be a human being and deserve all the rights and privileges that a non-cloned human has. A clone should not be a second-class citizen. It is speculated that clones would be considered as such.

The American Medical Association holds four points of reason why cloning should not take place. They are: there are unknown physical harms introduced by cloning, unknown psychosocial harms introduced by cloning, including violations of autonomy and privacy, impacts on familial and societal relations, and potential effects on the human gene pool. Serious ethical concerns arise by the future possibility of only harvesting organs from clones. Some people have considered the idea of growing organs separately from a human organism - in doing this; a new organ supply could be established without the moral implications of harvesting them from humans. Research is also being done on the idea of growing organs that are biologically acceptable to the human body inside of other organisms, such as pigs or cows, then transplanting them to humans, a form of Xeno-transplantation.

From a Latter-day Saint point of view, the Proclamation on the Family clearly does not agree with cloning. The Proclamation states: "We... declare that God has commanded that the sacred powers of procreation are to be employed only between man and woman, lawfully wedded as husband and wife. We declare the means by which mortal life is created to be divinely appointed. We affirm the sanctity of life and of its importance in God's plan." In other words, the power to create humans is only to be used in a marriage between husband and wife. Cloning only involves one parent, therefore it is not following God's plan in which a man's sperm and a woman's egg are needed to create life.

Cloning is the process of creating a copy of a biological entity. In genetics, cloning refers to the process of making an identical copy of the DNA of an organism. Cloning is an established process today, which holds the promise of repopulating endangered and even extinct animals.

Cloning can revolutionize the world and the way we live. Animal cloning has been one of the greatest frontiers scientists have conquered. However, there are various ethical and scientific issues related to cloning that have been debated. Foods from healthy cloned animals are deemed safe to eat. In recent years, there has been a shoot of new laws banning or regulating cloning around the world. In some countries, animal cloning is allowed, but not human cloning. Some advocacy groups are seeking to ban therapeutic cloning, even if this could potentially save people from many debilitating illnesses.

ANIMAL BREEDING BEHAVIOUR

Animal breeding is a branch of animal science that addresses the evaluation (using best linear unbiased prediction and other methods) of the genetic value (estimated breeding value, EBV) of domestic livestock. Selecting animals for breeding with superior EBV in growth rate, egg, meat, milk, or wool production, or have other desirable traits has revolutionized agricultural livestock production throughout the world. The scientific theory of animal breeding incorporates population genetics, quantitative genetics, statistics, and recently molecular genomics and is based on the pioneering work of Sewall Wright, Jay Lush, and Charles Henderson.

Breeding stock

Breeding stock is a group of animals used for purpose of planned breeding. When individuals are looking to breed animals, they look for certain valuable traits in purebred animals, or may intend to use some type of crossbreeding to produce a new type of stock with different, and presumably superior abilities in a given area of endeavor. For example, when breeding swine the "breeding stock should be sound, fast growing, muscular, lean, and reproductively efficient." The "subjective selection of breeding stock" in horses has led to many horse breeds with particular performance traits.

Purebred breeding

Mating animals of the same breed for maintaining such breed is referred to as purebred breeding. Opposite to the practice of mating animals of different breeds, purebred breeding aims to establish and maintain stable traits, that animals will pass to the next generation. By "breeding the best to the best," employing a certain degree of inbreeding, considerable culling, and selection for "superior" qualities, one could develop a bloodline or "breed" superior in certain respects to the original base stock.

Such animals can be recorded with a breed registry, the organisation that maintains pedigrees and/or stud books. The observable phenomenon of hybrid vigor stands in contrast to the notion of breed purity.

Backyard breeding

In the United States, a backyard breeder is someone who breeds animals, often without registration and with a focus on profit. In some cases the animals are inbred narrowly for looks with little regard to health. The term is considered derogatory. If a backyard dog breeder has a significant number of breeding animals, they become associated with puppy mills. Most puppy mills are licensed with the USDA.Chapter 8

Artificial Insemination

Artificial insemination (AI) is the deliberate introduction of semen into a female's vagina or oviduct for the purpose of achieving a pregnancy through fertilisation by means other than copulation. It is the medical alternative to sexual intercourse, or natural insemination.

Artificial insemination is a fertility treatment for humans, and is a common practice in the breeding of dairy cattle and pigs. Artificial insemination may employ assisted reproductive technology, donated sperm, and/or animal husbandry techniques.

Artificial insemination is the technique in which semen with living sperms is collected from the male and introduced into female reproductive tract at proper time with the help of instruments. This has been found to result in a normal offspring. In this process, the semen is inseminated into the female by placing a portion of it either in a collected or diluted form into the cervix or uterus by mechanical methods at the proper time and under most hygienic conditions. The first scientific research in artificial insemination of domestic animals was performed on dogs in 1780 by the Italian scientist, Lazanno Spalbanzani. His experiments proved that the fertilizing power reside in the spermatozoa and not in the liquid portion of semen. Few further studies under research station conditions helped this technique to be used commercially allover the world including India.

Artificial insemination is not merely a novel method of bringing about impregnation in females. Instead, it is a powerful tool mostly employed for livestock improvement. In artificial insemination the germplasm of the bulls of superior quality can be effectively utilized with the least regard for their location in far away places. By adoption of artificial insemination, there would be considerable reduction in both genital and non-genital diseases in the farm stock.

SYMPTOMS OF HEAT

The various symptoms of heat are

- The animal will be excited condition. The animal will be in restlessness and nervousness.
- The animal will be bellow frequency.
- The animal will reduce the intake of feed.
- Peculiar movement of limbo sacral region will b observed.
- The animals which are in heat will lick other animals and smelling other animals.

- The animals will try to mount other animals
- The animals will standstill when other animal try to mount.. This period is known as standing heat. This extends 14-16 hours.
- Frequent maturation (urination) will be observed.
- Clear mucous discharge will be seen from the vulva, sometimes it will be string like the mucous will be seen stick to the near the pasts of valva.
- Swelling of the valva will be seen.
- 11 Congestion and hyperemia of membrane.
- The tail will be in raised position.
- Milk production will be slightly decreased.
- On Palpation uterus will be turgid and the cervix will be opened.

A.I. Advantages - Disadvantages

Advantages and disadvantages:

Artificial insemination (A.I.) is deposition of semen into the female genital tract by means of instruments.

Advantages of artificial insemination

There are several advantages by artificial insemination over natural mating or servicing.

- There is no need of maintenance of breeding bull for a herd; hence the cost of maintenance of breeding bull is saved.
- It prevents the spread of certain diseases and sterility due to genital diseases.
- Eg: contagious abortion, vibriosis.
- By regular examination of semen after collection and frequent checking on fertility make early detection of interior males and better breeding efficiency is ensured.
- The progeny testing can be done at an early age.
- The semen of a desired size can be used even after the death of that particular sire.
- The semen collected can be taken to the urban areas or rural areas for insemination.
- 7 It makes possible the mating of animals with great differences in size without injury to either of the animal.
- It is helpful to inseminate the animals that are refuse to stands or accept the male at the time of oestrum.
- It helps in maintaining the accurate breeding and cawing records.
- It increases the rate of conception.
- It helps in better record keeping.
- Old, heavy and injured sires can be used.

Disadvantages of A.I

- Requires well-trained operations and special equipment.
- Requires more time than natural services.
- Necessitates the knowledge of the structure and function of reproduction on the part of operator.
- Improper cleaning of instruments and in sanitary conditions may lead to lower fertility.
- If the bull is not properly tested, the spreading of genital diseases will be increased.
- Market for bulls will be reduced, while that for superior bull is increased.

Semen Collection Methods And Evaluation

Various methods of collection of semen have been devised from time to time. The older unsatisfactory methods have gradually replaced by the new modern techniques.

There are three common methods.

1. Use of artificial vagina
2. By Electro-stimulation method.
3. By massaging the ampulae of the duct us differences through rectal wall.
4. The ideal method of semen collection is use of artificial vagina which is safe for sire and the collector also.

Artificial Vagina Method:

The artificial vagina has the following parts:

- A heavy hard rubber 2" lose, open at both ends with a nostle for air and water in and outlet.
- Inner sleeve of rubber or rubber liner.
- The semen receiving cone or rubber cone.
- Semen collection tube made of glass or plastic graduate in cc and its fraction correct to 0.1 CC
- Insulating bag Before using for semen collection all the parts are washed thoroughly and sterilized properly, and assembled as artificial vagina, the rubber liner is inserted into the hose; inverting both ends back by folding back from either side opening, and fastening with rubber bands. Now the space between the hard rubber hose and inner rubber liner forms a water tight compartment. The nostle at one end of the hose can be fixed.

PARTS OF ARTIFICIAL VAGINA

Turning through the threaded nut up or down. The water jacket of the Artificial -vagina is- filled with hot water at a temperature of 45 C (113 F) by opening the nostle. The graduated semen collection tube is fixed to the narrow end of the artificial vagina hose, and fastened by a rubber band. The inner side of the rubber liner on the anterior side of the artificial vagina is lubricated with sterile jelly to a length of 3 to 4 inches. Air is blown through the nostle into the water jacket, to create pressure in if, and the same is exerted the rubber linear, to simulate natural vagina.

The temperature of the artificial vagina is to be checked, at each collection, and it should simulate natural vagina at mounting time. If the artificial vagina is to mount later. If it is too cold ejaculate may not be there after a thrust, or even if ejaculate is there; it may be contaminated with urine, and becomes unfit for use.

Semen Collection Method. (A.v.)

The cow or dummy is secured in service create. The artificial vagina assembled is held at 45 angle from the direction of penis, and the thrust is that angle. The artificial vagina is held with the left hand by a right handed person; and when the bull mounts the cow, the sheath of the bull will be graphed by the operator, directing the gland penis into the artificial vagina, and then the bull gives a thrust to ejaculate. The operator should evince care so as not to touch the exposed past of the penis. After the bull dismounts, the artificial vagina is taken off from penis and the air vent is opened to release the pressure from the jacket. The water from the jacket is also drained by opening the nostle. This allows the ejaculate to flow from the cone to the semen collection tube. The semen collection tube is detached from the cone, plugged with cotton wool, and taken to the laboratory for examination. The rubber cone and the semen collection tube can be protected from external contamination or heat or higher, by covering with an insulation bag with zip.

Semen Storage

The discovery that bull semen could be successfully frozen and stored for indefinite periods has revolutionized AI in cattle. In 1949, British scientists discovered that addition of glycerol to the semen extender improved resistance of sperm to freezing. Glycerol acts to remove water from the sperm cell prior to freezing and prevents the formation of cellular ice crystals which would damage the sperm. There are two methods of freezing and storing semen: dry ice and alcohol (-100 degrees F) and liquid nitrogen (-320 degrees F). Liquid nitrogen is preferred because there is no evidence of fertility deterioration with age. Fertility gradually declines in semen stored in dry ice-alcohol.

Frozen semen can be stored indefinitely if proper temperature is maintained. A recent report told of a calf born from frozen semen stored for 16 years. Fresh, liquid semen can be successfully stored for 1 to 4 days at 40 degrees F. Semen is usually stored in glass ampoules. Other methods appear promising, particularly the French-straw. Several AI organizations have gone to this method exclusively. Artificial coloring is frequently added to semen extenders in order to distinguish one breed from another. Complete identification of the bull is required on each individual semen container.

Insemination Methods

There-are different methods insemination in different species of animals i.e. speculum method, vaginal method and recto vaginal method.

Recto Vaginal Method

In cattle the safe and best method of insemination is "Recto vaginal method of insemination". Cow which is in heat is well controlled placing it in a Travis. The inseminator will get ready by wearing a plastic apron, gumboots and gloves. The semen straw after thawing (keeping the semen straw in warm water for a minute to convert the freezed semen into liquid and the sperms become motile) is loaded in a sterilized A.I. gum and is covered with a plastic sheath. The inseminator will insert the gloved left hand into the rectum after applying the soft soap or other lubricant on the glove and back racked the animal, and the hand is further inserted and will catch hold the cervix through rectal wall. The A.I gum loaded with semen straw is passed.

Recto-vaginal method of insemination

Through the vulva to 'vagina and cervix and observed with the hand in rectum that the A. I gum reaches the cervix, then the semen is deposited by injecting the gun, and after depositing the semen the gun is removed, the empty straw and sheath are disordered.

Spectrum Method

In this method spectrum is placed in the vagina of the cow, which provides passage outside to the site of insemination, then inseminating tube is passed through the speculum and semen is deposited at the cervix insemination method.

Vaginal Method

Hand is passed through the vagina and the inseminating tube is guided by hand to the site of insemination and semen is deposited. Here there is a risk of contamination and injury of female genitalia.

FROZEN SEMEN AND STORAGE

Freezing of semen for successful preservation of spermatozoa, for long periods, is of great importance in livestock breeding and farm management. It has made it possible" to make available the use of outstanding proven sizes for larger number of cows, covering larger area, frozen semen shipment has become possible to different continents in the globe to any place connected with any service. Now a day if farmer wants to use of an outstanding size for inheritance of high milk yield, he can go in for frozen semen service provided his area is, covered by Artificial insemination, with supply of frozen semen.

At present frozen semen is used in most of the states in India. The technique of semen preservation in straws was developed in France. Freezing of semen is done with a special diluents, which has the following composition.

Sodium citrate dihydrate (angular) 2.4 y. 2.0 gm 8.0 ml 25.0%byvolume 50,000 units per 100 ml of semen Fructose Glycerol Egg Yolk Penicillin dilulent. Dihydro-streptomycin 50.0 mg per.100m1 of semen dilulent. Distilled water double glass distilled 100.Om1. The addition of glycerol to the dilulent makes the cells more resistant to the rigours of freezing and icy crystals, which form are smaller and smoother thus creating less damage to the spermatozoa. The addition of fructose to the diluent luprores sperm resistance to glycerol; and also provides nutrition.

Frozen semen is packed in single dose glass vials or plastic straws at +5 C. The final level of glycerol should be 7.0 to 7.6% during the freezing process. The antibiotics are added to inhibit bacteria and to kill pathetic organisms. The semen to be diluted in such a way that one ml. of extended semen will contain 20 million motile spermatozoa. The semen must be cooled carefully for spermatozoa to remain with life. The final temperature is lowered to -79 C or still lower. Quick freezing is done for a period of 3 to 5 minutes to -75 C with the help of atmosphere created by liquid nitrogen. In the slow freezing technique cooling is done at the rate of 1 C per minute from +5 C to -15 C. From -15 C to -31 C at the rate of 2 C per minute. From -31 C to 75 C at the rate of 4 to 5 C per minute. Thus taking 40 minutes in total, further cooling to -96 C can be done quickly as it is not critical after freezing. Before freezing the diluted semen in equilibreated for 3 to 5 hours or for the best 16 to 20, hours period in refrigerator at 5 C. Frozen semen facilitates the percent use of the semen diluted and frozen, and thus the delivery price is reduced, and it can be supplied with the gaps of months to the A.I technicians as against the supply of fluid semen every days or alternate days. Liquid nitrogen plays a vital role for storing the frozen semen straws, at a temperature of -196 C for longer periods.

Artificial insemination (AI) of cattle

Artificial insemination (AI) is the process of collecting sperm cells from a male animal and manually depositing them into the reproductive tract of a female. One can cite a number of potential benefits from the use of artificial insemination.

Increased efficiency of bull usage:

During natural breeding, a male will deposit much more semen than is theoretically needed to produce a pregnancy. In addition, natural breeding is physically stressful. Both of these factors limit the number of natural mating a male can make. However, collected semen can be diluted and extended to create hundreds of doses from a single ejaculate. Also, semen can be easily transported; allowing multiple females in different geographical locations to be inseminated simultaneously, and semen can be stored for long periods of time, meaning that males can produce offspring long after their natural reproductive lives end.

Increased potential for genetic selection:

Because artificial insemination allows males to produce more offspring, fewer males are needed. Therefore, one can choose only the few best males for use as parents, increasing the selection intensity. Furthermore, because males can have more offspring, their offspring can be used in a progeny test program to more accurately evaluate the genetic value of the male. Finally, individual farmers can use artificial insemination to increase the genetic pool with which his or her animals can be mated, potentially decreasing effects of inbreeding.

Decreased costs:

Male animals often grow to be larger than females and can consume relatively larger amounts of feed. Also, male animals are often more strong, powerful, and potentially ill-mannered and thus require special housing and handling equipment.

Increased safety for animals and farmers:

As mentioned, male animals can become large and aggressive. These factors mean that maintaining a bull on a farm may be dangerous. Also, because of the relatively larger size of adult males than females, natural mating is more likely to result accidents and injury to either the cow or the bull than is artificial insemination.

Reduced disease transmission:

Natural mating allows for the transfer of venereal diseases between males and females. Some pathogens can be transmitted in semen through artificial insemination, but the collection process allows for the screening of disease agents. Collected semen is also routinely checked for quality, which can help avoid problems associated with male infertility.

Artificial insemination has some potential drawbacks, however, that must be considered. First, it can be more laborious. Male animals instinctively detect the females that are in the correct status for conception. With artificial insemination the detection work falls on the responsibility of the farmer. Poor detection results in decreased rates of fertility. Also, increasing the number of offspring per male has selective advantages only if the best males can be accurately determined. Otherwise this process only decreases the genetic variability in a population. Increasing the number of offspring per male always reduces the gene pool. The benefits of more intense selection must be balanced against the negative effects of decreased variation.

ARTIFICIAL INSEMINATION TECHNIQUES

The technique of inseminating a cow is a skill requiring adequate knowledge, experience and patience. Improper AI techniques can negate all other efforts to obtain conception. Semen must be deposited within the tract of the cow at the best location and at the best time to obtain acceptable conception rates. Early methods of AI involved deposition of the semen in the vagina, as would occur in natural mating. Those methods are not satisfactory. Fertility is low and greater numbers of sperm are required. Another method which gained popularity was the "speculum" method. This method is easily learned, but proper cleaning and sterilizing of the equipment is necessary, making it more impractical to inseminate than with the rectovaginal technique which is the most widely used AI method today.

In the recto-vaginal technique a sterile, disposable catheter containing the thawed semen is inserted into the vagina and then guided into the cervix by means of a gloved hand in the rectum. The inseminating catheter is passed through the spiral folds of the cow's cervix into the uterus. Part of the semen is deposited just inside the uterus and the remainder in the cervix as the catheter is withdrawn. Expulsion of the semen should be accomplished slowly and deliberately to avoid excessive sperm losses in the catheter. The body of the uterus is short; therefore, care should be taken not to penetrate too deeply which might cause physical injury. In animals previously inseminated, the catheter should not be forced through the cervix since pregnancy is a possibility. Since research data show little variation in conception rates when semen is placed in the cervix, uterine body or uterine horns, some people recommend incomplete penetration of the cervical canal and deposition of semen in the cervix.

The recto-vaginal technique is more difficult to learn and practice is essential for acceptable proficiency but the advantages make this method of insemination more desirable than other known methods. With practice, the skillful technician soon learns to thread the cervix over the catheter with ease. If disposable catheters are used and proper sanitation measures are followed, there is little chance of infection being carried from one cow to another.

Timing of Insemination for Maximum Conception

A frequent question concerning AI is: What time during estrus should cows be bred for greatest chance of conception? Since estrus may last from 10 to 25 hours there is considerable latitude in possible time of insemination. Much research work has been conducted on this subject.

Controlled investigations were conducted by Trim Berger and Davis at Nebraska in 1943. These and other studies show that conception rate is lower when cows are bred prior to mid estrus or later than 6 hours after cessation of estrus (standing heat in this case). Maximal conception is obtained when cows are inseminated between mid estrus and the end of standing estrus, with good results up to 6 hours after estrus.

Success in insemination timing is dependent upon a good heat detection program. In large herds, this means assigning individual responsibility for heat detection and a continued education program for labor. A successful heat detection program and subsequent proper timing of insemination will pay dividends in increasing reproductive efficiency.

6

Learning Behaviour in Animal

Many animals live out their lives following fixed and apparently unvarying routines. Among numerous species of solitary insects, for example, the life cycle consists of the following unvarying events: the females lay their eggs on a particular plant or captured prey; the newly hatched larvae immediately start eating and then follow a standard sequence of developmental stages; the adults recognize appropriate mates by a set of fixed signs, perform a fixed sequence of mating responses, provision their eggs with suitable nourishment, and finally die before the next generation hatches. The same unchanging sequence is repeated generation after generation. And it is, of course, eminently successful. The same set of responses is invariably elicited by the same set of stimuli, because those responses were, and continue to be, adaptive. Where circumstances do not change, there is little need for an animal's behaviour to change. Even many aspects of the behaviour of mammals show a similar fixity. A dog withdraws its foot if it is pricked and a young child his hand if burned; both people and rabbits blink whenever an object is moved rapidly toward their eyes; the feeding behaviour of young infants of virtually all mammalian species consists of sucking elicited by contact with the lips.

SIMILAR TOPICS

Whenever the same response is always appropriate in a particular circumstance, there is little reason why an animal should need to learn what to do in that circumstance. But the world is not always so stable a place. The food supply that was plentiful yesterday may be exhausted today, and the foraging animal that always returns to the same spot will starve to death. Moreover, a particular food supply may be temporarily depleted but will be replenished if left long enough; the successful forager needs to remember where the supply was and when it was last visited, so as to time a return to advantage. In other words, circumstances may change, and the same response is not always appropriate to the same stimuli. Knowing what behaviour is appropriate may depend, therefore, on keeping track of past events.

Possible Explanations of Behavioral Changes

Related Topics

- Animal social behaviour

If an animal's behaviour toward a particular stimulus changes, one must look for an explanation of that change. One possible explanation is that the change is due to learning, but there are numerous other possibilities. If a definition of learning is to be provided, that definition must specify when to attribute the change to learning, and when to other causes.

At least two other major causes of behavioral change have been widely recognized. The first of these is motivation. A laboratory rat may pick up, chew, and swallow a pellet of food at one moment; half an hour later, after having eaten 20 grams of food, the rat will simply ignore any further pellets offered. Similarly, a male rat may mount and copulate with a receptive female introduced into his cage, but he will not repeat this pattern of behaviour endlessly even if offered the opportunity to do so. Some male territorial birds, such as chaffinches, will feed amicably beside other males at certain times of day or certain seasons of the year, but at other times they will launch an attack on any intruding male. In all these cases, it is more reasonable to attribute the change in behaviour not to anything the animal has learned but rather to a change in the creature's motivational state.

It should not be thought, however, that just because all of these examples can be attributed to a single item-i.e., motivation-that their detailed explanation will always be the same. The analysis of motivation is itself a large field of study, and it has proved to be more profitable to concentrate on the specific explanation of individual cases of changes in behaviour rather than to search for broad explanatory principles that end up being nearly vacuous. Nonetheless, it does seem possible to draw a contrast between motivational explanations for such changes and those that appeal to learning.

- Observational learning

A second broad class of changes in behaviour can be attributed to maturation. We are inclined to ascribe the unfolding pattern of behaviour that emerges over the first few weeks of life to this ill-defined process. Newborn rat pups, for example, are relatively helpless; their eyes do not open for about two weeks, and their main sources of sensory input are probably touch and smell. As their sensory apparatus matures, the pups exhibit changed behavioral responses. The other obvious instance of a maturational change in behaviour is that which comes with sexual maturity: sexually mature adults of most species behave toward one another in ways quite different from those of younger members of the species. It is not only courtship and mating behaviour that change with sexual maturity; for instance, male puppies urinate in the same way as females, by squatting, and it is the onset of sexual maturity that produces the adult pattern of cocking a hind leg.

The concept of maturation is probably no better defined than that of motivation, and it is equally important to stress that it must cover a number of different processes. And, as with motivation, it is more profitable to analyze each case in detail, in order to uncover the precise mechanisms involved, than it is simply to label a change as an example of maturation. Indeed, at the level of physiological process, it seems probable that both motivational and maturational changes are often due to alterations in the hormonal state of the animal, and the distinction between the two is largely one between the unidirectional nature of the change in the case of maturation contrasted with the cyclical change common to short-term motivational states.

But how are these changes discerned from those that might be ascribed to learning? In many cases, of course, the answer is because a precise causal explanation has been provided: a great deal is known, at a physiological level, about the changes in brain and body associated with the motivational states of hunger and thirst. Even without any such detailed knowledge of the underlying mechanisms, it is possible to insist that certain changes in behaviour be attributed to motivation rather than to learning if the opportunity to learn anything relevant was lacking and the opportunity for a motivational change was present. If, for example, an animal that has been deprived of food for a long time behaves in one way toward food-related stimuli, but some hours later, after having been given ample opportunity to eat, it behaves differently toward those stimuli, the obvious interpretation is a motivational one. This interpretation would be strengthened if the animal had not come into contact with these stimuli during the intervening period and had been given, as far as one could judge, no other opportunity to learn anything about them. Learning, in other words, depends on certain kinds of opportunity, and a definition of learning may well turn out to be no more than a specification of the particular set of opportunities and experiences that produce it.

CIRCUMSTANCES THAT PRODUCE LEARNING

Britannica Stories

A particular change in behaviour is attributed to learning, then, because it is possible to specify the set of circumstances that produced it. What are those circumstances? It is common to claim that learning depends on practice. (An older generation of experimental psychologists would have claimed that it depended on "reinforced" practice.) This definition can be misleading, however, if it causes one to attribute to learning all behavioral changes that follow what appears to be practice. In other words, it is not enough to show that an animal appeared to engage in practice and its behaviour subsequently changed. A temporal correlation of this sort does not establish a causal connection. Young birds, for example, are unable to fly, and their first attempts at flight are clumsy and ill-coordinated. Casual observation suggests that young birds improve with practice, gradually perfecting the set of skills they display as adults, but experimental analysis suggests that this practice may be unnecessary. Young birds have been brought up under restricted conditions that completely prevented their flying. When released at the age at which normally reared birds fly proficiently, the experimental subjects flew-without practice-as successfully as those that had spent their time in trial flight. The development of the skill appears to depend more on the maturation of strength and agility than on specific practice.

Types of Learning

Simple Nonassociative Learning

When experimental psychologists speak of nonassociative learning, they are referring to those instances in which an animal's behaviour toward a stimulus changes in the absence of any apparent associated stimulus or event (such as a reward or punishment). Studies have identified two major forms of simple nonassociative learning, which are to some extent mirror images of one another: habituation and sensitization.

Habityuation

A classic example of habituation is the following observation on the snail Helix albolabris. If the snail is moving along a wooden surface, it will immediately withdraw into its shell if the experimenter taps on the surface. It emerges after a pause, only to withdraw again if the tap is repeated. But continued repetition of the same tapping at regular intervals elicits a briefer and more perfunctory withdrawal response. Eventually, the stimulus, which initially elicited a clear-cut, immediate response, has no detectable effect on the snail's behaviour. Habituation has occurred.

Habituation can be defined in behavioral terms as a decline in responding to a repeatedly presented stimulus. As such, it is a very widespread phenomenon, one that can be observed in animals ranging from single-celled protozoans to humans. Most animals behave differently to novel and familiar stimuli: the former sometimes elicit startle responses, sometimes investigatory or exploratory responses; the latter often apparently are ignored. The suggestion that habituation is a simple form of learning, however, implies that it can be distinguished from some even simpler potential causes of this sort of change in behaviour. One reason why an animal might stop responding to a stimulus is that it no longer detects the stimulus; i.e., some form of sensory adaptationmight have occurred. Another potential cause is fatigue: perhaps some temporary refractory state is produced by repeated elicitation of the same response, making it impossible to perform that response again. Whether or not one would want to call either of these processes a form of learning is doubtful. But both behavioral and physiological evidence establishes that habituation cannot be explained in these terms.

Sim

The critical behavioral evidence is that habituation can be disrupted by almost any change in the experimental conditions. If repeated presentation of one stimulus leads to habituation of a response, the same response can still be elicited by a different stimulus. Even if the experimenter presents a novel stimulus that does not itself elicit the response in question, its presentation may restore the response on the next trial in which the originally habituated stimulus is presented. This latter observation, usually referred to as an instance of dishabituation, seems to rule out any simple sensory adaptation; both observations rule out simple effector fatigue.

Neurophysiological analysis of habituation in various mollusks-for example, in the sea snailAplysia-has confirmed that habituation need not depend on changes in the activity of sensory or motor neurons. In the case of Aplysia, researchers have studied the gill withdrawal reflex, a response that rapidly habituates to repeated stimulation of the snail's siphon or mantle shelf. But habituation still occurs even if it is elicited by direct, electrical stimulation of the motor nerve, bypassing the sensory receptors completely; and recording from the sensory nerve during normal habituation reveals no decline in its level of activity. These observations eliminate sensory adaptation as a possible cause of the animal's having ceased to respond to the stimulus. Effector fatigue can be ruled out by showing that direct stimulation of the motor neurons controlling the withdrawal response can still elicit a perfectly normal reaction even after the response has completely habituated. Research shows that habituation in Aplysia depends on changes in the activity of more central neurons.

Repeated tactile stimulation of the siphon, leading to habituation of the withdrawal response, causes changes in the activity of the motor neurons innervating the response. Specifically, these motor neurons show a decline in excitatory postsynaptic potential, which is the electrical change that enables the nerve impulse to cross the gap (synaptic cleft) that separates one neuron in the pathway from the next. The decline in excitatory postsynaptic potential short-circuits the response. Moreover, the presentation of a novel stimulus, sufficient to dishabituate the behavioral response, restores the postsynaptic potential.

BRITANNICA STORIES

Habituation occurs even in animals without a central nervous system-probably in single-celled protozoans; certainly in animals such as the coelenterate Hydra, which have a diffuse nerve netand do not appear to be capable of associative learning. Among mammals, habituation of certain reflex responses can be observed even in "spinal" subjects, that is, those whose spinal cord has been severed from the brain. There can be little doubt, then, that habituation is not only widespread, but that it also can be a relatively simple phenomenon. There is, however, no guarantee that it is the same phenomenon wherever it appears. The waning response to a repeatedly presented stimulus admits of a number of different explanations. In principle, as we have already seen, it might be due to sensory adaptation, effector fatigue, or a more central neural change. These distinctions make rather little sense in the case of a single-celled animal. And one should not necessarily expect the habituation observed in a spinal mammal to involve precisely the same mechanisms as those responsible for comparable behavioral effects in an intact animal. Some psychologists have proposed theories of habituation that appeal to processes of classical conditioning. Such a theory is not likely to apply to the habituation observed in an animal that shows no capacity for classical conditioning.

Habituation is usually, as here, classified as an instance of simple, nonassociative learning. It is supposedly nonassociative because all that happens in the course of habituation is that a stimulus is repeatedly presented and the animal's behaviour changes; there is, on the face of it, no other event with which the stimulus can be associated. Habituation must therefore, it appears, be understood by reference to some change in the pathway between stimulus and response, and the work with Aplysia and other mollusks shows how this analysis may proceed at the physiological level. But if habituation is not always the same phenomenon, it is possible that different processes may underlie the habituation of the startle response to a loud noise in an intact mammal. And despite appearances to the contrary, those processes may involve some associative learning. One suggestion is that novel stimuli elicit a biphasic response: an initial increase in startle responses, which include components of emotion or anxiety, followed by a rebound in the opposite direction. Habituation occurs when the latter, rebound response becomes conditioned to the stimulus, occurring sooner and sooner with each repetition of the stimulus and thereby damping down and eventually canceling out the initial reaction. An alternative possibility is that long-term habituation depends on associating the repeatedly presented stimulus with the contextin which it occurs, a suggestion that would explain why presentation of the stimulus in a different context sometimes leads to dishabituation.

Britanninca Lists & Quizzes

The generality of habituation implies that this behavioral phenomenon has considerable adaptive significance; if true, it would be quite reasonable to expect that a number of different mechanisms might have evolved to produce the behavioral result. The adaptive value of habituation is not difficult to see. A novel stimulus may signify danger, and an animal should react to this stimulus either by withdrawing or at least by orienting toward it to see what will happen next. But if the same stimulus occurs again with no further consequence, it is probably safe: regular repetition of the same stimulus implies that it is part of the background, such as the waving of a branch in the wind or the shadow caused by a piece of seaweed floating with the waves. If the stimulus is not dangerous, time should not be wasted on it. Withdrawal, especially in the case of a snail into its shell, is a time-consuming effort, incompatible with such vital activities as searching for food. If it is important, therefore, for animals to be wary of novel stimuli, it is equally important that they should discriminate the novel and potentially dangerous from the familiar and probably safe.

Sensitization

The effect of habituation is to eliminate unnecessary responses, but the main function of learning has usually been thought to be the production of new responses. Traditional psychological theories of learning have assumed that the learning of new patterns of behaviour comes about through the association of a new response with a particular stimulus. Consequently, psychologists usually have either ignored the possibility that nonassociative processes might be sufficient to increase the probability of a new response or regarded it as a nuisance that interferes with the measurement of associative changes. They have rarely treated it as a subject worthy of study in its own right.

Connect with Britannica

This is unfortunate, for the nonassociative phenomenon of sensitization is probably fairly widespread, and it provides a simple means of acquiring adaptive behaviour. Sensitization is said to occur when the repeated presentation of a particular significant stimulus (such as food or electric shock) lowers the threshold for the elicitation of appropriate behaviour to the point where a second stimulus, not normally capable of calling forth that behaviour, now does so. A typical example is provided by the behaviour of the marine worm Nereis. If the worm is kept in a small tube and fed at regular intervals, it becomes progressively more likely to respond to any novel stimulus, such as a change in illumination, by exploratory, food-seeking movements toward the open end of the tube. If, on the other hand, the worm receives mild electric shocks at regular intervals, it becomes progressively more likely to respond to a novel stimulus by withdrawal.

The first point to note about sensitization is its relationship to habituation. Habituation refers to a decline in the probability of responding to a repeatedly presented stimulus. Sensitization, by contrast, refers to an increase in the probability that behaviour appropriate to a repeatedly presented stimulus will occur, even in response to another stimulus. Although these two outcomes cannot be observed simultaneously, it is quite possible that the same operation-repeated presentation of a stimulus-can simultaneously engage two different processes, one causing a decline in the probability of responding to that stimulus, the other causing an increase. Experimental analysis suggests that both processes are real and may be engaged in the same experiment, so that the observed change in behaviour actually results from a mixture of the two. Typically, the process of habituation wins out, and what is observed is an overall decline in responding. But a common finding in habituation experiments is that responding initially increases before declining; the implication is that the initial presentations of a stimulus result in more sensitization than habituation, while further presentations produce more habituation than sensitization. A second factor influencing the relative importance of the two processes is the intensity or significance of the stimulus. A weak stimulus, or one with little intrinsic biological significance, will show relatively rapid habituation and little or no initial sensitization. A stronger stimulus, especially one, such as food or shock, that has substantial significance to the animal, may show marked sensitization and relatively little habituation.

Trending Topics

The second point about sensitization is that it may mimic the effect of associative learning or conditioning. As has been mentioned, in a classical conditioning experiment a neutral stimulus, such as a change in illumination, is paired with the delivery of a significant stimulus, such as food or shock. Repeated pairing causes the neutral stimulus to elicit responses initially called forth by the significant stimulus; for example, a change in illumination that has been associated with an electric shock would come to elicit retreat or withdrawal. But in the case of the worm Nereis, experiments demonstrate that the light would come to elicit this change in behaviour whether or not it had been paired with shock: all that is needed is sufficient exposure to the shock. To attribute the change in behaviour toward the light to its association with food or shock, one must show that this change is greater than that which would have resulted from sensitization alone.

The physiological processes underlying sensitization, like those underlying habituation, have been analyzed in experiments on such invertebrate species as Aplysia. Not surprisingly, the mechanisms involved appear to mirror one another. Whereas habituation is correlated with a decline in postsynaptic potentials, sensitization is correlated with an increase in the magnitude of postsynaptic potentials at the same locus.

Although sensitization has often been treated as a nuisance whose effects must be controlled in studies of habituation or associative learning, it remains a process worthy of study in its own right, for the behavioral changes it produces can have significant adaptive value. Without requiring the presumably more complex neural machinery necessary to subserve associative learning, sensitization enables animals to respond to local variations in the occurrence of significant events. If an animal's sources of food tend to occur together (that is, they are not distributed randomly in time or space), then it pays that animal, having once found food, to continue to behave in a food-gathering manner. Conversely, the animal that is increasingly wary after exposure to danger will have a better chance of evading a lurking predator. Sensitization thus enables an animal to take advantage of statistical regularities in the occurrence of significant events, without requiring it to detect other events that predict the significant ones. No doubt, further advantage accrues to the animal that can perform such calculations, for associative learning provides a powerful means of predicting the future. But there can be equally little doubt that such a process requires a more elaborate nervous system.

ASSOCIATIVE LEARNING: CONDITIONING

The study of animal learning in the laboratory has long been dominated by experiments on conditioning. This domination has been resisted by critics, who complain that conditioning experiments are narrow, artificial, and trivial, and, as such, miss the point of what animals are adapted to learn. From the critics' point of view, one unfortunate effect of their attacks has been the progressive refinement and elaboration of the theory of conditioning to the point where it can often explain the exceptions to which they drew attention. This is not to insist that associative learning is the sole, or even the most important, form of learning in vertebrates, but rather to introduce the idea that the processes underlying conditioning may be more interesting than older theories and an earlier generation of textbooks suggested.

The notion that learning depends on practice also seems unduly restrictive and is, perhaps, an unnecessary legacy of an earlier version of behaviourism. It is not obvious that an animal should actually have to engage in a particular form of behaviour in order that this pattern of behaviour should be affected by learning. In many cases, indeed, no such practice seems necessary. The young of many songbirds must, it is quite clear, learn their species-typical song. There are several aspects to this learning process, one of which may indeed involve practicing the song at the beginning of the young bird's second season. But another critical aspect is simply exposure to the adult song at some point during the autumn of the young bird's first year, at a time when the young bird does not practice singing at all. Deprived of such experience, chaffinches and song sparrows produce an extremely impoverished version of the adult song; some finches may develop a song more characteristic of another species if that is what they heard during this period of their life. There are numerous other examples where learning appears to depend more on the opportunity to observe than on the opportunity for practice.

This suggests that the definition of learning will have to refer to changes in behaviour that are attributable to particular kinds of experience. The danger now is that, as with motivation and maturation, the definition of learning will be so broad and vague as to be useless. As in those cases, it may be more profitable to concentrate on more detailed analysis of particular instances of learning. Such analysis has, for example, led to widespread agreement on the definition of classical conditioning, a particular type of learning whose study was pioneered by the Russian physiologist Ivan Petrovich Pavlov. In a typical experiment on classical conditioning, an experimenter might arrange a correlation between the ringing of a bell and the delivery of food to an animal. The animal predictably learns to direct food-related activity toward the sound of the bell. Analyses of such experiments have led to the definition of classical conditioning as a type of learning that occurs when there is a correlation between two stimuli and the animal's behaviour toward one of these stimuli changes in a predictable manner determined by the nature of the other. This definition, which will be expanded later in this article, is useful because it specifies both the circumstances responsible for learning (a temporal correlation between two stimuli) and the general way in which experience of those circumstances changes behaviour (the animal starts directing toward one stimulus responses that are related to those normally directed toward the other). Experimental psychologists and ethologists, however, have devised a tremendous range of procedures for studying learning in animals. The range and variety are such that it may be well-nigh impossible to formulate a meaningful definition of the circumstances that produce learning, for the definition either will be so restrictive that it clearly applies to only a fraction of the cases that should be regarded as instances of learning, or it will be so broad that it says nothing.

Rather than pursue any further the attempt to find an all-embracing, single definition of learning, it seems more useful to provide narrower definitions for particular cases, along the lines suggested above for classical conditioning. One consequence of this approach is that it may encourage the belief that learning consists of a large number of distinct processes that have nothing in common with one another. It is, of course, an open question as to whether this is true: it is certainly possible that, just as with the concept of motivation, the layman's concept of learning encompasses a large number of different cases whose underlying mechanisms are quite distinct. It is important not to prejudge this issue. Insistence on a single, global definition may well tend toward just such prejudgment by encouraging the belief that learning is a single, common process. To start by drawing some distinctions between types of learning does not rule out the possibility of seeing whether the various cases studied do have anything in common.

In the final analysis, as is true of all scientific definitions, the definition of learning is a matter of theory. It has been said that a good scientific definition is the end product of good theory and experiment, not the starting point. Thus, there is a single process of learning if it turns out to be possible to devise a single theory that adequately accounts for the variety of cases in which learning is assumed to occur. Superficial appearances may be deceptive: just because the circumstances that produce learning in two cases, along with the consequences of that learning, appear quite different, it does not follow that the processes underlying learning are different. For instance, the phenomenon of filial imprinting, first seriously analyzed by the Austrian ethologist Konrad Lorenz, appears to be a highly specialized form of learning in which a newborn animal (e.g., a chick, duckling, or gosling) rapidly learns to follow the first salient, moving object it sees. Normally this object will be the mother, but Lorenz discovered that the range of potential imprinting objects is large, extending from Lorenz himself to a bright red ball. There is no question but that some process of learning occurs here, and Lorenz assumed it to be highly specialized. Yet one theory seeks to explain imprinting in terms of simple classical conditioning. Whether or not the account of imprinting provided by this theory is correct, the point is made that how learning is defined and whether it is defined as a single, monolithic process or as many specialized processes are, in the end, questions of theory.

Learning

Learning is characterized by persistent and measurable changes in behavior which are not associated with fatigue, altered motivation, or maturation. Some information or knowledge is acquired and is then used to alter the individuals actions and responses. Learning as an adaptive behavior allows individuals to adapt to specific environment challenges.

Non-Associative Learning

The mode of learning THAT develops in the absence of its association with any reinforcement is called non-associative learning. It is of following two types: Habituation and Sensitization: Habituation refers to a gradual decrease in behavioral responses with repeated encounters of a particular stimulus, which proves of no consequence. It depends on a change in the synapse between the sensory and motor neuron. Independent of conscious motivation or awareness it aids in distinguishing novel and meaningful information from the general background noise. Sensory Adaptation refers to a change in the responsiveness of a sensory system when confronted with a constant stimulus.

In contrast, sensitization refers to an increase in behavioral responses following repeated applications of a particular stimulus. Following sensitization, very little stimulation is then required to produce exceedingly large effects. Initial. light stimulation of peripheral skin receptors may not activate nociceptive (i.e., pain) neurons of the spinal cord. Continued, repeated stimulation of the same pathways will bring about central sensitization suggesting the presence of an irritating or potential more damaging skin issue. Tissue damage or continued inflammation may thereby cause chronic pain conditions. In long-term potentiation (LTP), physiological effects of subsequent synaptic signals are strengthened following initial activation. This process is thought to form an integral component of memory and learning. Repeated stimulation of different neural centers can also stengthen coordinated neuronal firing and thereby lead to seizures.

Associative Learning

The ability to learn is a basic foundation of intelligence. Humans are particularly keen observers of the world around them and are thereby able to reliably identify the presence of different regularities and generalizations. Patterns that are of consequence to our daily lives are readily learned, which suggests the presence of powerful biases in their search and representation. We tend to remember information more easily if we have uncovered it ourselves than if they are simply presented to us.

An effort to form mental images out of any material to be learned may often enhance retention. Dual Code Theory suggests that most information can be committed to memory in either verbal/linguistic code or mental image code. The use of multiple codes for harnessing any material thus may enhance performance by increasing the number of retrieval paths.

Classical Conditioning

A form of associative learning, Classical conditioning requires an unconditional reflex, where an unconditional stimulus (US) elicits an automatic, unlearned (unconditional) response (UR). If a neutral stimulus (NS) tends to precede it, an association is made and the conditional response (CR) becomes transferred onto the (previously neutral) conditional strimulus (CS); a conditional reflex has been learned. For instance, food (US) elicits salivation (UR) in a dog as a natural response. If the sound of a bell (NS) frequently occurs before the food (US) is presented then the mere sound of the bell (CS) will elicit salivation (CR). A mistranslation of "conditional" as "conditioned" meant that in English the CS and CR were referred to as conditioned stimulus and conditioned response, and the verb "to condition" was derived to refer to the process responsible for the establishment of new CR. (From: The Dictionary of Ethology and Animal Learning Edited by Rom Harre and Roger Lamb, The MIT Press, 1986)

Fear conditioning allows organisms to acquire affective responses, such as fear, in situations where a particular context or stimulus is predictably elicits fear via an aversive context (e.g., a shock, loud noise, or unpleasant odor). Little Albert Experiment

Conditioned Taste Aversion (i.e., Garcia conditioning) occurs when a subject experiences symptoms of a toxic, spoiled, or poisonous substance such as nausea, sickness, or vomiting after consuming unfamiliar food. Learned aversion of such taste is an adaptive trait that trains the body to stay away from items that are likely poisonous (e.g., berries or mushroom). This association sometimes occurs in subjects even when sickness was merely coincidental and not related to the food.

OPERANT CONDITIONING

Operant conditioning, sometimes called instrumental conditioning or instrumental learning, was first extensively studied by Edward L. Thorndike (1874-1949). Thorndike's most famous work investigated the behavior of cats trying to escape from various home-made puzzle boxes. When first constrained in the boxes the cats took a long time to escape from each. With experience however, ineffective responses occurred less frequently and successful responses occurred more quickly enabling the cats to escape in less and less time over successive trials.

B.F. Skinner (1904-1990) extended the theories proposed by Thorndike about 40 years after Thorndike published his works. His ideas of animals operating on the environment, lead him to analyze how behavior is changed by its consequences. The contraption that became known has the Skinner box measured if a task was completed and the duration of time it took for the task to be completed. The Skinner box consisted of a bar on the wall that when pressed triggered the release of a food pellet. It was Skinners belief that by rewarding an animal when an appropriate action occurred it would increase the likelihood that the behavior would be repeated. When the rats, placed in the box, accidentally tapped on the bar on the wall a pellet was released and Skinner observed the amount of time it took the rat to find the pellet. As the rat learned that a pellet released each time he happened to step on the bar, the rat learned to press the bar and to immediately find the food. This training is what became known as operant conditioning. Reinforcement always strengthens the behavior that preceded it and may consist of any outcome that the animal considers desirable. Reinforcers can broadly be categorized into positive reinforcers (i.e., obtaining access to something the animal wants) or negative reinforcers (i.e., removing something considered unpleasant or painful). Punishment is always designed to discourage further performance of the behavior it is paired with. Positive punishment achieves that by inflicting an unwanted outcome such as unpleasant sounds or odors, electric shocks, or pain, while negative punishment removes access to something the animal wants. Skinners operant conditioning is seen in the learning behavior of many animals as well as humans. Oftentimes it goes completely unnoticed, but it can also be used as a deliberate tool in training. As young children explore their world and learn how to eliminate unpleasant stimuli and satisfy desires through proper behavior they are learning through operant conditioning.

Operant conditioning is used extensively in training animals towards performing quite complex tasks. Shaping involves the step-wise reinforcement of successive approximations. This training method relies on the animal to exhibit an interest in showing a wide range of behaviors as well as in obtaining rewards. When a desired behavior is performed, the trainer rewards this action. As this method relies on compliance from the animal, most suited to this training are a combination of positive reinforcers (e.g., a small item of food) for behavior compatible with the trainer's expectations, and negative punishment (e.g., withholding that food item) for behavior that is not. Initially rewards are issued for even a crude approximation or a component of the behavior. While these attempts may have earned the animal a reward at the beginning, subsequent rewards are contingent on progressively narrowing the behavior to what the trainer aims for.

Imprinting

Latent Learning

Latent learning refers to an individual's ability to learn associations without explicit reinforcement. Exploratory behavior serves to acquire an understanding of the spatial relationships of objects. It includes the formation of "cognitive maps" of the surrounding. Daily success depends on knowldge of spatial relationships. Hummingbirds and bees recall the location and status of harvesting of flower resources, and several species of birds are able to track a large number of seed stores.

OBSERVATIONAL LEARNING

Bandura illustrated the power of an instructional strategy in which children are allowed to observe behavior of individuals who are significant to them. Subsequently they are likely to imitate their role models especially when such behavior is reinforced. His work illustrates that observers readily adapt their behavior based on what they see around them, even when they received no encouragement or incentives to do so. Integrating a continuous interaction between behaviors, cognitions, and the environment, his social learning theory stresses the importance of observational learning, imitation, and modeling.

Language Learning

Human Language

Humans show exceptional skill to communicate with fellow conspecific using verbal and gestural symbols. Such competence is arguably one of mankind's greatest assets as well as a key to most of our species' achievements. It brings us together with our peers and enriches us through an exchange of experiences, thoughts and value systems. It endows us with the means to acquire skills for situations that we may have never personally encountered before. A staggering range of regionally distinct languages and dialects, grouped in larger language families, not only serves as a system of communication but also tags us for membership within a specific group. In addition to bringing us together, it has thereby been at the heart of an ancient source of division. Although few things in biology ever group cleanly into one of the nature vs. nurture extremes, this particular division seems to be purely cultural. Regardless of the specific race, ethnic or regional group that we happen to have joined through birth, we all are able to acquire competence in any human language. In particular, we seem to acquire our native language without formal education, long before many of our intellectual capabilities have mnatured, and simply by immersion into the particular language environment of parents and relatives during the first few years of our life. The scientific study of the nature and structure of languages is called linguistics.

Universal Grammar. BF Skinner (1957) suggested that infants learn language through a process described as operant conditioning, namely, via the monitoring and management of reward contingencies. Skinner's position would be that a four-term contingency analysis comprised of motivating operations, discriminative stimuli, responses and reinforcing stimuli would be the means by which behavior could be explained. In children and infants this process would be expanded by what he called "shaping", "prompting" and other stimuli modeling, imitation and reinforcing procedures. Language acquisition then is a process that would take thousands of instances of such training, and this appears largely to be what takes place. Critics who do not understand the inductive power of this approach largely assert that this view now "appears quite simplistic." However, this argument to complexity is not dissimilar to arguments against Darwin's theory of natural selection. How could the amazing complexity of the animal kingdom come about through such a simple means? Surely, we would need something more complex to explain the variety of animals? Darwin's theory of selection is now accepted as such a means, despite it's "apparent simplicity". Similarly the mechanism of operant conditioning can be seen as sufficient to account for very complex forms of behavior in a wide variety of circumstances without appealing to unproven, non-data driven speculative theoretical approaches like those of Noam Chomsky and others.

Language learning is clearly the most complex task any of us will ever undertake. Yet, the process appears to be considerably less painful than acquiring an understanding of calculus, or organic chemistry. Noam Chomsky (1975) has long argued that this paradox is best explained by the view that humans, and in particular children, have innate abilities that support the acquisition of a language. It is clear that we seem to be naturally good at it, especially before we reach puberty. Moreover, we appear to have a natural need to fill our world with language; in the absence of formal language tutoring a form of language structure develops anyway. Some say a specialized language faculty seems to aid in this process, one that includes innately specified constraints on what forms are possible. These innate, language-specific, information processing mechanisms may be encapsulated in language module of the brain. However, these innate faculties are inferred, hypothesized explanations with no foundation in fact. No biological location has been found for them, no genetic location, no brain structure. It is all inference.

All human languages, even spontaneous ones, show many common principles of language acquisition as well as rules of grammar. The concept of universal grammar proposes that this is due to a set of innate rules, which guide how we acquire language and how we construct valid sentences in it. It thus attempts to explain language in general, and not simply describe the construction of any one specific language per se.

There are many alternative theories of human language consistency besides the speculative theory of "universal grammar". One theory is that human environments possess common structures and human language simply responds to the commonality of the world. Universal grammar is a speculative, unproven hypothesis that is still awaiting confirmation and has no evidence to support it other than "rational argument".

Language Acquisition. At very young age, we acquire our native language by listening to, guessing at the meaning of, and imitating the symbols used by tutors around us. Moreover, during these early years we learn to walk and talk without any explicit need for understanding how we are doing what we are doing. In this process we seem to be helped by a set of Inherent learning strategies, the ability for optimized pattern perception of common, ambient symbols.

Infants are exceptionally broad in their abilities to perceive sound qualities. In fact, as infants we can distinguish many more language sounds than we can as adults. During the first year of life, infant brains are actively engaged in optimizing acoustic perception for the language sounds that surround them. Such early acquisition of information about native language depends on perceptually mapping both the critical aspects of language, and statistical properties of speech.

It is now clear that infants perceive the various phonetic units, track the frequency of different formants, and extract the boundaries of words from running speech. Patricia Kuhl suggests that language acquisition is based on a combination of factors to provide a powerful discovery procedure for language. Evidence suggests that initial perception parses speech in a universal way in all human infants. Infants have inherent perceptual biases that segment phonetic units without providing innate descriptions of them. They were able to parse and discriminate a wide range of basic phonetic units. Adults, in contrast, are only able to discriminated phonetic units that occur in their first language, but fail to distinguish those that are not used there. Japanese adults, for example, fail to discriminate phonetic boundaries of r vs. l, boundaries that do not exist in Japanese. Such discrimination is based on general auditory processing mechanisms, rather than on innate phonetic feature detectors for speech. Language learning requires mapping these probabilistic patterns into language strategies. As infants detect frequency patterns in language input they identify higher-order units. Infants thus discover the critical parameters and phonetic dimensions of the sounds used in their native language. Sensory processing becomes optimized by experience for enhance perception of the specific language around them. Vocal learning unifies language perception and production where vocal learning depends on a comparison of one's own vocalizations to those of others. Imitation forms the integral bong between the perception and production of language abilities and together they become optimized for the first language. If a second language is learned later on, it will carry the accent typical for the speech motor patterns of their primary language, even following long-term instruction. Similarities in infant-directed speaking styles (increased pitch and exaggerated stress) enhances language learning by assisting infants in discriminating phonetic units, as well as by capturing attention.

Broca's Area underlies the ability to produce speech, but it is not critical for understanding language. Patients with damage will fail to form words properly, and speech is halting and slurred. Wermicke's Area is essential for the ability to understand language. Patients with damage to this area can speak clearly, but the words make no sense (i.e., word salad). The Arcuate Fasciculus connects these two areas. Damage to this connection causes conduction aphasia where language is understood, but neither can words be repeated, nor does own speech make any sense. Capabilities for speech are not distributed evenly across the two halfs of the brain. Speech is only disrupted when amobarbital is selectively used to anesthetize only the half of the brain which contains these speech centers. Imaging techniques (e.g.

fMRI) have identified that bilingual individuals utilize an overlapping set of neurons in the language areas for these two languages. In contrast, individuals who have acquired a second language later in life will likely rely on separate neuronal areas in these speech centers. Late bilingual speakers are also less likely to show strong lateralization of speech function. This suggests that as two language systems are learned together early-on, they can share the same brain centers without causing catastrophic interference. In adult learning, the best sites of brain real estate have already been taken up by the first language, thus, any new language learning must coopt 'new' territory adjacent to it or on the other half of the brain.

Bird Song

Birds communicate information about danger, food, sex, group movements and many other purposes via acoutic signals. A subset of these have been termed song, as they frequently feature with extended, tonal, melodic characteristics. The Zebra Finch's song, for instance, includes several introductory notes followed by a string of syllables within an extended melodious pattern. Sonograms (i.e., a plot of the intensity of pitch against time) are commonly used as a primary tool for studying and comparing bird songs.

Respiratory muscles force streams of air from large air sacs through the bronchi. Membranes in the syrinx vibrate as air expressed from bronchi passes over them. Syrinx muscles for left and right sound producing structures can act independently, and many birds are able to sing harmonies with themselves. Song appears to play a role in advertising for sex and species recognition as song complexity frequently coincides with the presence of ornate plumage. It also stimulates and synchronizes courtship behavior, stimulates reproductive readiness in females, and contributes to pair bond maintenance. Local song dialects exist in many species.

Successful song in most adult male songbirds depends on memorizing the calls of a conspecific tutor during an earlier, sensitive phase in life. The appropriate song repertoire is acquired in a series of distinct stages. Young birds, during an early Sensory Phase, listen to a conspecific tutor and thereby obtain information about the characteristics of its own song. Only a very specific subset of surrounding songs is actually accepted as suitable, suggesting the presence of an in-built song template. Following this sensory phase, young birds actively begin to vocalize themselves. Their Subsong is an atonal, noisy, meaningless repetition of sounds, which lacks recognizable syllables. Akin to human Babblingbirds practice coordinated movements of the respiratory system, sound producing organs, and related structures (e.g., tongue). During Sensory-motor Phase, young birds spontaneously produce Plastic Song, consisting of vocalizations with distinct syllables and recognizable elements. Such "work in progress" will include elements from the song of tutors and elaborate them into a variety of syllables and phrases that even exceed what eventually will be used in its adult song. The ability to hear its own vocalizations are critical for normal development. In transition to the Mature phase, birds adopt a Crystallized Song with syllables and syntax structure that is characteristic of its species. Once established, these song patterns remain fixed in many species, are no more disrupted by deafening, and are presented intact during each subsequent breeding season. In contrast open-ended learners (e.g. starlings and canaries) retain the capacity to adjust or alter their song throughout life.

Song production is under the control of multiple hormonal systems from embryonic gonads. Injections of testosterone induce adult males to sing, even out of season, while similar injections in females have no such effect. The presence of estrogen during male development appears to be essential. When estrogen is blocked experimentally in developing males, testosterone injections fail to elicit song. However, when estrogen had been delivered to developing females, injection of testosterone elicited song in them (Konishi).

Several neural centers with a role in song have been identified. The Higher Vocal Center (HVc) is a group of neurons in the forebrain that is larger in (singing) males than in (non-singing) females. Damage to it blocks song production in adults. The nucleus of the archistriatum (RA) in males is larger than in females and its neurons increase in size and dendritic arborization during song learning. Damage to this area blocks song production in adults. The lateral magnocellular nucleus of the anterior neostriatum (LMAN) is neither sexually dimorphic nor shows seasonal change in neuron size or number. Its ablation in young birds interferes with song acquisition but its ablation in adults brings about few deficits as long as song had already been been learned prior to damage. Area X of the paraolfactory lobe (Area X) is sexually dimorphic and new neurons are added in song learning. Damage to it interferes with song acquisition in young birds but not in adults.

As in every behavioral system, a series of independent questions can be addressed for song behavior in Zebra finches (Taeniopygia guttata).

- Proximate Causation: Zebra finch song production requires the flow of air through semi-independent vibrators in syrinx and vocal tract. The presence of song, and song repertoire size are reflected in sexual dimorphism of its controlling brain areas and nuclei. A learning pathwayesists separate of a motor pathway. Singing, which is largely restricted to males, is under the control of androgens.
- Ultimate Causation: Song in Zebra finches is a learned vocalization used during courtship and defense of a territory. Advertising the individual's presence it serves to elicit mating opportunities from females and to stimulate the partner's reproductive behavior and physiology. Moreover, it functions to claim a territory and to repel competitors from it.

- Phylogeny: Virtually all 9000 species of birds have the ability to vocalize, including crows, turkeys, owls or nightingales. A large subset of them, including the zebra finch, are characterized by complex vocal organs, distinctive brain circuitry for song, and acquisition of species-characteristic vocalizations through learning. Taxonomically these are all restricted to a single order - the Passeriformes.
- Ontogeny: The emergence of adult zebra finch song illustrates the interactions of genetic and environmental factors in behavioral development. After periods of listening to the songs of tutors, starting its own partial vocalizations, rehearsing and adaptating its own song, the species-specific adult version slowly emerges. Song circuits exhibit extensive plasticity even in adults with ongoing neurogenesis and seasonal changes in neuronal morphology.

White-crowned sparrows (Zonotrichia leucophrys nuttalli) males sing a single song that shows considerable geographic variation in the form of stable dialects. Bilingual and blended strategies exist at the boundaries. The distinctiveness of the song depends on patterns of natal dispersal and the timing of learning. Subject to reinforcement by the song of neighbors, the system is highly dependent on auditory feedback. The work by Peter Marler, Doug Nelson and others for over 30 years illustrates how genetic and environmental factors interact during the development of a complex communication system. Males generally establish territories during late plastic song with a repertoire that consists of ~4 different songs. Improvisations yield individual-specific songs which closely match that of the nearest rival.

Cross-fostering experiments illustrate the role of auditory templates in song learning. Young birds reared in the presence of taped song will learn and present that song, even if the tape came from another species. A Song sparrow raised with a swamp sparrow tape will experience little difficulty to learn the swamp sparrow song. Birds in Isolation experiments are raised without access to intact adult song (i.e., no template) and will subsequently show deficiencies in their own song upon maturation. The song does nonetheless contain valid elements of intact adult song. Moreover, Deafening experiments, which deafen birds at hatching, results in song that still contains some valid elements but is an even cruder version than those of isolated birds. When Song preference experiments present young birds with a wide range of conspecific and heterospecific songs, they recognize and preferentially learn the song of its own species. Birds raised in Mixed syllables experiments in the presence of a mixture of swamp- and song sparrow syllables, will accurately produce these syllables in their song but lack the normal adult syntax.

Castration Experiments have shed light on the roles of hormones in song learning. Swamp sparrows that are castrated early in development have low testosterone levels compared to their male siblings. They acquire song but progress to plastic phase only. Treating such birds with injections of testosterone (Enhanced Testosterone Experiment) immediately crystallizes the song. Interfering with testosterone function in adult birds (Reduced Testosterone Experiment) degrades previously crystallized song back to plastic.

Brown-headed cowbirds (Molothrus ater) are gregarious birds that follows cattle herds. Brood parasites that are raised by parents of different species, no consistent, conspecific tutor available. So, how do they learn their own conspecific song? One of the strongest stimuli is the bird's own crystallized song and feedback from females is important (i.e., action-based learning).

PARALLELS BETWEEN BIRD SONG AND HUMAN SPEECH LEARNING

To learn their language, humans and white crowned sparrows follow similar steps, recognition, practice, and clarity. The best time for a human to learn their language is from toddler to twelve years old (Macdonald) and the best time for a white crowned sparrow to learn its song is between 10-50 days from when they were born. Both genes and environment play roles in determining the way each communicate.

The first step for humans and white crowned species in learning their language can be considered recognition. White crowned sparrows and humans listen to a tutor before they begin to communicate their language. The tutor teaches the language and the particular dialect, according to the area. Just like humans, the sparrows also have a dialect depending on where they live. A white crowned sparrow has certain genes that only allow them to learn its own species' songs. Experiments have been performed in which the sparrow was kept in isolation and played tapes of other song sparrows, the white crowned sparrow will not imitate the other species' song, but will sing an odd song unlike either species' song. Humans also cannot communicate with each other using a different species' language. A study was done in Avignon, France that observed children who were brought up by wolves. It was observed that the children spoke no language at all. Before studies of this kind were done it was thought that the children might learn to communicate with the wolves, just like the Tarzan story. This is not true, the children could not speak the human language nor the language of the wolves (Macdonald). It is a combination of environment and genes that tell white crowned sparrows and humans to learn their proper language. Genes tell the species to only learn the language of their own species and environment plays a factor in determining which dialect each will use.

Once the white crowned sparrow or infant has recognized its species, song, they can begin to practice the language. Infants first communicate by making sounds. The sounds can range from grunts to sounds that mimic their surrounding environments. For example, before babies can say words, they might say moo- moo when looking at a cow. After sounds come words. By the time the infant reaches one year, they should be making sentences out of words (Macdonald). The white crowned sparrow also does not immediately master its song. The sparrow will first sing a short subsong derived from the tutor's full song. The sparrows keep practicing their subsong just as infants practice their words. After the sparrow has mastered its subsong it can start to form a full song. In both species it is necessary for them to hear themselves in order to vocalize their language correctly.

In conclusion, both white crowned sparrows first listen and recognize their particular species' language. A white crowned sparrow has a tutor to teach the song. Infants usually have parents that teach them their verbal language. Then they must practice the language, starting off slow with sounds and building up. Both species will continue to improve and clarify their language throughout their lives.

Motor Learning

Coordinated movement requires that motor neuron innervate one or more muscle cells with finely tuned neural patterning. For a person to perform even the simplest motor task, the activity of thousands of these motor units must be coordinated. It appears that the body handles this challenge by organizing motor units into modules of units whose activity is correlated. The cerebellum and basal ganglia are critical for motor learning. Motor learning is "relatively permanent", as the capability to respond appropriately is acquired and retained.

Memory

Memory, in its broadest sense, refers to an individual's ability to acquire information. The formation of memory traces is a complex task which all organisms appear to be able to do. Fundamentally, such an ability requires a series of distinct steps: Information must be encoded as meaningful associations are assessed, the resulting information must be stored in some form, and the memory record must be retrieved when needed. Humans exhibit little conscious awareness of mental processes, such as memory. These were thus long considered inaccessible to objective, scientific study. Inferences about the course of encoding, consolidation and storage of memory thus mostly rely on indirect methods such as testing for retrieval abilities. Consolidation of memory appears to involve independent paths at different time frames. Such evidence derives from the existence of different critical time frames for its formation, and its susceptibility to disruption with different pharmacological tools.

Sensory Memory

Sensory memory provides a means for organisms to remember the experience of a particular sensory perception after the stimulus itself has stopped. It thereby provides an immediate 'snapshot' of an animal's sensory experience. Sensory memory formation is involuntary and does not depend on attention. A different type of sensory memory is stored for each particular modality - Iconic memory is a fast decaying store of visual information that briefly stores an image which has been perceived for a small duration; Echoic memory is a fast decaying store of auditory information; Haptic memory is a type of sensory memory that represents touch stimuli. Details that prove significant are transferred to more permanent and conscious components of working memory. For example, a toddler who is burnt transfers information from tactile sensory memory to short term memory due to the significance of the stimulus.

Short-Term Memory

Short-term memory allows the recall of something from several seconds to as long as a minute. Its strength does appear to depend primarily on attention and not rehearsal. It is thus highly vulnerable to disruption when attention shifts elsewhere. The amount of information that can be held is on the order of 4-6 numbers. This amount can be boosted by grouping them into distinct chunks as for telephone numbers (e.g., 202-456-1414). The ability to recall such information is contingent on transient patterns of neuronal activity in regions of the frontal and the parietal lobe. Memory of written language may primarily rely on acoustic components rather than visual ones, as we are much more likely to make mistakes between letters that sound similar (E,P,D) than those that have visual similarities (O, Q)

Long-Term Memory

This type of memory, lasting hours to months, critically depends on a transfer of the information from short term memory using repeated rehearsal. The hippocampus appears to be an essential structure in such routing. Sleep is thought to improve the consolidation of information, possibly by hippocampal replaying of activity from the previous wake period. Electrical activation of hippocampal circuits reportedly are linked to feelings of Deja Vu.

Declarative (Explicit) Memory

Declarative memory refers to the ability to become conscious of, or declare, facts and experiences. It is also referred to as Explicit memory, when it involves direct recalling information that had been obtained from the external world. Representing knowledge of standard textbook material or events, it is best formed by actively recalling the material in spaced intervals. Compared to other forms or memory it is more volative; more easily formed and more easily forgotten. The primary neural basis appears to reside in the medial temporal lobe. Bilateral damage to this area results in anterograde amnesia, as in the famous case of the anonymus memory-impaired patient HM. A surgical procedure for epilepsy left him with damage to his brain in the medial temporal lobe on both sides. the hippocampal formation, parahippocampal gyrus, the entorhinal cortex, and the amygdala. He subsequently suffered from severe anterograde amnesia, where transfer of new events into long-term memory was impaired. He was unable to recall events once his attention had focused elsewhere. Semantic Memory is the ability to consciously recall knowledge of facts that are independent of a specific time and place. medial temporal lobe, diencephalon. Episodic Memory refers to the ability to explicitly recall information about a specific event that has occurred at a specific time and place, medial temporal lobe, diencephalon

PROCEDURAL (IMPLICIT) MEMORY

Implicit memory, which does not need to involve conscious awareness in the act of recollection. Procedural Memory regards the learning of motor skills and habits. Formation requires repeated reinforcement, repetition and practice over many trials rather than recollection. Once formed it is less likely to be forgotten. It is also less easily transferred to related tasks than declarative learning. Striatum, basal ganglia, Deficits can be assessed using a serial reaction time (SRT) task, backwards reading, mirror drawing, probabilistic classification, artificial grammar learning, or prototype abstraction. Motor Responses with Classical Conditioning, Cerebellum. Emotional Responses with Fear conditioning involves an organism's ability to acquire fear responses to a previously neutral stimulus. This occurs when it becomes paired with an aversive stimulus, such as a shock or loud noise. James Watson's Little Albert experiment illustrated that children learned fear of objects when encounters with them were paired with loud noises. Priming refers to a context-dependent activations of clusters of neocortical neurons. As they become more activated, they are more likely to come into consciousness. Even reflex pathways are capable of surprising plasticity. For example, circuits that control motor patterns for walking will be subjected to optimization with sensory input.

Patient HM suffered severe memory impairment following bilateral damage to his medial temporal lobes, hippocampal formation, parahippocampal gyrus, entorhinal cortex, and amygdala. His short-term memory remained intact, he was able to largely recall past events, and his ability to learn new motor skills was not disrupted. However, he suffered from severe anterograde amnesia, i.e., the ability to commit new events to long-term memory.

Memory mechainsms critically depend on changes in synaptic functioning. Hebb Synapses and their role in classical conditioning. If a synapse succeeds at driving a postsynaptic neuron above threshold, its subsequent effectiveness is strengthened.

Long-term potentiation is based on changes in neural signals which will potentiate a neural response for 1-2 weeks. potentiation involves increased release of the excitatory neurotransmitter glutamate.

Molecular

During habituation, repeated stimulation of a sensory neuron leads to a smaller activation of the postsynaptic motor neuron. The primary cause for this is a progressive reduction of Ca inflow into the presynaptic terminal, decreased transmitter release, and a smaller activation of the postsynaptic target. In sensitization as a result of electrical head shock, release of serotonin activates second messenger systems and phosphorylation of key target molecules. Long-Term Memory depends on new protein synthesis and the formation of new synaptic connections.

Mechanisms of Animals and Microorganisms

Different animal species follow different mechanisms to survive in extremely low or high temperature. These mechanisms are physiological which aids to rise body temperature in cold environment or to lower body temperature in high temperature environment. When it comes to increasing body temperature in cold atmosphere, the shivering is one of the well known techniques seen in most of the animal species. This is a mechanism which contracts the muscles and hence generates the heat. The second mechanism is decrease in the flow of blood to skin; this hinders the blood from actually cooling down before reaching heart, brain and other muscles and organs. Fluffing out the fur to increase the insulation is the third mechanism which is good for majority of the mammals except humans. (R. C. Tinsley in 2002)

Behavioral regulation mechanism is observed in Antarctic penguins. While the temperature drops to -22F. The one month old penguin chick normally gets insulated against this low temperature hence many other penguin chicks huddle it tightly and act like a large insulated organism. This specific process is so effective that penguin cluster has to move often to stop ice from melting down. The reptiles and amphibians in temperature below zero degrees Celsius burrow themselves underground or sometimes find other locations for shelter.

Some fogs even survive through the lower underground temperature which is up to -40 C. These animals generally stock their blood with antifreeze chemicals and glycerol right at the beginning of the winter. Though wood frogs freeze at this low temperature they reduce the damage by certain mechanisms. These frogs withdraw almost all the fluids from their blood vessels and organs and store it in extracellular spaces. Hence as a result ice crystals can expand without the need of tearing cells and blood vessels. Moreover these frogs have chemicals which cause formation of ice crystals gradually instead of chunks. The amazing blood clotting attribute of these frogs can instantly repair ruptured blood vessels.

These physiological mechanisms which defend body temperature of the animal depend on brain's hypothalamus in and near area. The anterior hypothalamus regulates the body temperature. Suppose an experimenter/ researcher heats this area of animal's brain the animal body will start sweating even in the cooler environment. Suppose the same area is been cooled then the animal will start shivering even at the warmest room.

MECHANISMS FOR SURVIVING EXTREME HIGH TEMPERATURE IN ANIMALS

Corrective Mechanism for Temperature Control

The animals are capable of regulating temperature through negative feedback control. Once again hypothalamus acts like a receptor in temperature regulation by identifying the temperature fluctuation. Skin of the animal plays a primary role in detecting external environment temperature. This information gets conveyed to hypothalamus which in turn transmits the nerve pulses in order to achieve corrective mechanism. There are various survival techniques which enable animals even in the hottest days.

Excessive sweating is one of the corrective responses which are aimed to bring down the animal body temperature. In Vasodilation the blood vessels which are close to animal's skin gradually become highly dilated which means that the larger surface area is created for heat of the environment to be lost right from blood vessels which carry heated blood. The drop in metabolic rate is another type of mechanism for balancing the body temperature observed in animals.

For example a gemsbok (normally seen in hottest deserts of South Africa) has sparkling coat which can reflect sunlight which in turn aids in reducing the body temperature. The gemsbok has a network of exclusive veins on its face that cools down the blood before it reaches the brain via a process of thermal panting. This process lets the gemsbock's temperature to move upward. The same process is also followed by ostriches. In addition the features of ostrich insulate it from the rising heat. The ostrich has the capacity to cool itself down by running all around in a sunny day, the fanning out its feathers aids the loss of conventional heat. The body heat gets carried away due to air movement via its features.(Fatik Baran Mandal in 2007)

Some of the animals living in deserts develop the ways which can dissipate the excessive heat of the environment. Some of the well known physiological adaptations are long year or long appendages etc. Some animals are paler in skin color. This feature in fact serves two major functions like absorbing the minimal heat which helps in keeping the animal cooler. Camouflaging according to the desert surrounding helps these from predators. The black vulture excretes urine on its legs which indeed helps in cooling down. The same mechanism is followed by the turkey.

Survival of Mechanism in Plants at High Temperature:

Similar to animals even plants comply various survival methods which help in surviving even at the very high or very low temperature. For example desert plants make use of both behavioral and physical mechanisms in order to adapt to the high heat and the aridity of the desert. Xerophytes are the types of plants which alter the physical structure to adapt to the increasing temperature of the external environment. A cactus is one of the examples of xerophytes. These plants have unique means of conserving water. The limited number of leaves or no leaves in such plants brings down the transpiration level. The plants which grow very long roots in order to acquire moisture near or at the water table are known as Phreatophytes. (David A, in 2002)

Some of the behavioral mechanisms followed by desert perennials to survive in extremely high temperature is keeping themselves dormant. These plants spring into life when once more the temperature stabilizes. The cactus adaptation of rose family is known as one of the highly drought resistant plants on earth. The shallow root system, limited leaves, capacity to store water on the stems or on spines for waxy and shade skin in order to steal moisture are some of the traits of these plants to adapt to the enhancing temperature. The other varieties of cactus like beavertail cactus, jumping cholla and more.

The photosynthesis process of cacti is based on the chlorophyll available on the external tissue of the plant skin and its stems. The plant spines in fact shade it from hot sun and aids in collecting the moisture. The extremely shallow root systems of such plants are normally radial. This is because of the fact that they can store the water on the stems core as well as on roots core. Hence these plants can easily survive dry and high temperature climate and survive just from the water been collected from mere one time rainfall.

Eliminating leaves is one of the best methods followed by trees or shrubs of deserts which replace these leaves with thorns and hence eliminates transpiration. These types of plants generally have green bark and trunks which is capable of producing food and the seal in moisture.

Survival Mechanism in Plants at low Temperature:

Winter is one of the most difficult seasons for plants which bring number various problems due to low temperature in external environment. In winter season as the number of daylight hours is lesser than that of night hours, the green plants will not be capable to photosynthesize and produce the food. Moreover in winter season when the temperature drops lower than zero the water becomes highly inaccessible to these plants. Hence the survival technique in such environment becomes mandatory. Some plants can survive easily through the winter by entering to a resting stage.(Peter D. Moore, Richard Garratt in 2008)

Indeed the experiments have shown that extremely low temperature slows down the plant growth and also stops the circulation of water in the plant sap. But yet, there are various plants which are accustomed to different mechanism to adapt to the surrounding low temperature environment. In few plants known as annual plants reproduce in summer and die in the last days of summer. The seeds produces will survive through the winter season. The seeds gradually start germinating and grow into upcoming spring. This type of plant is known as annual plant.

Few plants survive via their underground storage tuber/organs. For example potatoes, carrots, tap roots, onions and more. Autumn and the summer is the season when these plants start storing the food in underground storage organs. At the time of winter, food reserves will be utilized for growing the new leaves and then the shoots over the ground. Such plants are known perennials. Apart from the food storage capacity these plants can also produce huge number of seeds every year.

Some plants add new roots and branches to their body every year as an act of survival mechanism in low temperature. Having their branches and trunks on ground in all the winter exposed to freezing winds and snow these plants will have to survive the entire winter. The carrot plant can produce storage organs in the foremost year of its life and there comes the flower and then the seeds in the second year. Such plants are known as biennial plants, which follow the survival mechanism based on the stage of its life. Thus all in all the plants too have inbuilt survival mechanisms which help them survive even at the lowest temperature and thus survive up to the life span.

Survival of Mechanism in Microorganisms at Hing Temperature:

There are various microorganisms in the world, which can survive even in the toughest environment under very low or high temperature. For example few bacteria are adaptable and versatile enough to get accustomed to the varying environment temperature. Some bacteria exist in different sizes and shapes. These shapes may include long like rods, round like cocci and some in coiled shape. Besides, these shapes mentioned there are various types and shapes of such microorganisms. These creatures can exist even under wacky conditions and such types of organisms are known as extremophiles. The extremophiles microorganisms can survive in extremely low temperature as well as extremely high temperature. These can easily survive in environment with temperature up to 121°C. (A. K. Pavlov, V. N. Shelegedin, in 2003)

These organisms grow in too acidic ph value which is generally less than 3.0. The Ferroplasma acidiphilum and archaeobacteria are seen in the Iron mountain mines. This type of resilent mine water cells comprise copper, zinc and iron. These microorganisms survive by releasing the hydrogen in the form of a waste and by oxidizing iron. Research has proved that Ferroplasma Acidiphilum not only uses iron just to extract energy but in fact these also use it in the form of an element for cellular proteins. Researchers have also discovered that proteins of these microorganisms comprise iron atoms in almost all of its proteins. These atoms of irons are utilized as anchors in order to hold the protein chains.

Thermophile is a bacteria which normally grows at a very high temperature from 50 degree Celsius to 80 degree Celsius. These bacteria are categorized into facultative and obligate thermophiles. The thermophiles known as hyperthermophiles undergoes the optimal growth only above the temperature of 80 degree Celsius.

One of the highly remarkable discovery was to find the fact that Thermas Aquaticus which is an Archaea microorganism is capable to resist high temperature. This is due to the survival mechanism found in Thermus Aquaticus which makes it capable enough to duplicate DNA at a very high temperature. The Vibrio vlunificus is one of the curved bacterium which can survive marine environments. These bacterium's growth rate is very high in warm seawater. The Deinococcus radiodurans are another type of microorganisms which can easily survive in the toughest environment with high level of exposure to radiation. These organisms can survive in thousand times radiation atmosphere than a common person. These organisms have the capacity to put its genome with perfect fidelity. This is one of the highly remarkable features of these microorganisms because despite the radiation shatters the DNA the bacterium proteins can easily repair them. The extra genomes let the bacterium to regain one complete set of genome after it gets exposed to the radiation.

The microorganisms like Archeans incorporate inhabitants of few of the extreme environment on the earth. Some of them can survive right near rift vents inside the deep sea at more than 100°C, some can. live in highly hot springs, or also in highly acid or alkaline waters. Few of the archaeans can also survive desiccating impacts of high temperature. The very sensitive and light pigment bacteriorhodopsin brings the Halobacterium(group of Archaea) chemical energy and the color. Bacteriorhodopsin has purple color and pumps the protons into the membrane. These protons are later used in the ATP synthesis which in fact is the main energy source of each cell. Archeans are those microorganisms which can easily survive in extreme habitats like thermal vents.

There are several microorganisms which can stand the extreme low temperature of external environment and can go on with its day to day activities without much of impact. For example Psychrophiles are the microorganisms, capable of exhibiting optimum growth even at 15 degree Celsius or lower than that. Some of the other species collected from extremely low temperature environment which include Psychrobacter sp, Arthrobacter sp, members of Genera Halomonas, Sphingomonas and Pseudomonas also exhibit the same properties at low temperature. The Psychrobacter species are the unique microorganisms which can reproduce even at the temperature of -10 degree Celsius to -40 degree Celsius. Bacteria proteins are main source which make these microbes to live as well as reproduce even at the lowest temperature. (Andrew Bell, John Swenson-Wright, Karin Tybjerg in 2008)

The survival and the growth of low temperature is in fact also relevant to storage of food or to food processing, stress responses and normal mechanisms of these bacterial survival. Moreover, these adaptations which Psychrobacter has created to inhabit the permafrost actually enables hypothesis regarding the possible microbial life even in the extraterrestrial environments.

All in all the inbuilt features and traits of these microorganism or plants or animals help in surviving at the extreme temperature easily without hindering its normal activities.

DISTINGUISHING MEN FROM ANIMALS WITH CULTURE

"For culture is the sociological term for learned behaviour: behaviour which in man is not given at birth, which is not determined by his germ cells as the behaviour of wasps or the social ants, but must be learned anew from grown people by each new generation. The degree to which human achievements are dependent on this kind of learned behaviour is man's great claim to superiority over all the rest of creation; he has been properly called 'the culture being animal." --- Benedict

Different people may have different definition of culture. The most common definition of culture is that culture means the mode of human activities, such as human knowledge, learning and belief which are integrated in the symbolic thoughts. Culture includes ideas about identity, nature, social relationship and so on, as well as artifacts. This topic is based on several assumptions. Firstly, this essay assumes that humans are not animals and only 'men' have culture while animals do not have any. I also assumed that 'culture' can be definite.

In the following, I will discuss the perspective of human in the point of view for zoologists, and also the inter-specific relationships between human and animals. This might counter my argument of 'culture distinguishes men from animals'. Secondly, I will discuss

Zoologist accounts on the human-animal's inter-specific relationship

To zoologists, all animals (including human beings) are equally interesting. For them, all animals undergo the inter-specific relationship which is in the food chain, i.e. prey, symbiont, competitor, parasite and predator, and human beings exist in these pair of species. According to The Naked Ape (1967) by British zoologist, Desmond Morris, "Any species which competes with us for food or space, or interferes with the efficient running of our lives, is ruthlessly eliminated Large carnivores have been out other serious competitors and these too have been eliminated wherever the population density of out species has risen above a certain level." Same as in the predator relationship, we humans are involved in it. Morris (1967) stated that larger carnivores such as sharks which are the killers are accounted for more humans' deaths than any other is one cannot devour the nutritious corpuses it produces.

Moreover, Morris (1967) also pointed out that the inter-specific relationship appears to human and other animals not only in the food chain, but also in other ways. For example, human beings could not resist to say 'Hi' or any other greetings to his/her own pet, such as his/her own dog. Even we all know the animals will not understand, we still could not resist the temptation. We could not explain that. The response was just triggered off inside us when we see the animal. Furthermore, Morris (1967) also pointed out that out love or hatred to the animals is also included in the human-animal relationship. Through this, we can see the inter-specific relationship between human beings and animals exist in the vast amount of cultures, or even, we carried further in these relationships than other animals. Therefore, to zoologists, humans are not unique in these respects.

Biological differences between human and other animals

However, biologist anthropologists and other biologists may argue that there is something which is different in human than in other animals, which this will lead to human are not animals (which I assumed before). So what distinguishes us from other species on animals? Gorillas and chimpanzees are close primate relative to humans. Compare to chimpanzees, human has 98 percent of genes same as the chimps. The differences between human and chimpanzees were brought by the 2 percent difference in our genes. Biological anthropologists believe that human beings have got a familiar skeletal hallmark. The major difference is that humans are bipedalism (standing on two legs) and the adaptation of the straight, upright gaits. Humans are having particular small teeth and large braincase as well. The 2 percent difference also brought the difference of using modern language to communicate idea. Human beings also use culture, which are complex ideas to survive. These are all the anatomical character of human beings.

Human evolution in human brains: What is in the Homo sapiens mind?

According to Jerome Kagan (2004), at the end of the first year, human and chimpanzee infants are similar to each other, "Both of the species locomote, attend to unexpected or unfamiliar events, and remember where an attractive object disappeared ten seconds earlier." However, 24 months later, maturational changes occur in human brain and it brought four unique quantities for humans: (1) Children can infer varied thoughts and feelings in others; (2) Children use a symbolic language with a grammar and semantic categories for events that share no physical feature; (3) Understand for the concepts of good. Bad, right and wrong; (4) Become consciously aware of some of their intentions and feelings. Moreover, Kagan also stated that "chimpanzees are incapable of appreciating the partner intends to engage in reciprocally cooperative act has no implication for the gaining of food or protectionâ Every two-year-old child makes this simple assumption automatically." Therefore, it is believed that a pair of chimpanzees could not throw a ball back and forth between them, while a pair of humans will assume to throw the ball back.

One of the unique characteristic in human brains is the feeling of guilty. Human is the only animal species which will feel guilty, it is not possible even in our close primate relatives such as the chimpanzees. Kagan explained that "Guilt requires an agent to know that a voluntary act that could have been suppressed has hurt another. Guilt requires the ability to reflect on a past action that injured another in some way, to realize that the behaviour could have been inhibited, and to appreciate that the self was the cause of the ethical violation." During human evolution, human developed to be conscious and aware to other's feeling, while other animals did not. Thus, the feeling of guilty is one feature which makes human unique.

Kagan pointed out another unique quality of Homo sapiens is the consciousness of one's feelings, intentions and properties, which is called sensory awareness. According to Kagan, there is not just only one form to awareness. It is believed that all these awareness require the brain circuits. There is awareness in properties such as consciousness in one's physical features, beliefs, talents and social categories. The other form of awareness is the awareness that one is about implement or suppress an action. Neuroscientist Michael Gazzaniga believes that there is another form of consciousness which is the interpretation of one's feelings and perceptions. However, these four forms of consciousness are not evolved in the same time, they evolve in different ages in the humans' early childhood. In the first year, there will be the appearance of the awareness of sensory. In the second year, children will smile after completing a difficult task. It is not until the third year of life that children will be aware of their intentions and that they can describe what they are doing. Finally, in the fourth year, children can regularly integrate the present moment with their recollections of the past and begin to be able to interpret other's feeling and perceptions. However, for apes, they do not possess any other forms of consciousness apart from the sensory consciousness. In terms of consciousness, human is unique.

Another unique quality in the Homo sapiens' is that they are the animal species which are willing to limit their offspring from inclusive fitness (the ability of an organism to ensure the survival of their own offspring or to see how their genes pass on). The family size in human families is thus decreasing. Meanwhile, in other species, they will reproduce the offspring as much as they can, so that their species would not extinct due to any environmental factors, such as drought, lack of food, lack of shelter, disease and so on. For example, female fish will lay their eggs as much as possible so that there will be a greater chance for the eggs being fertilized. Moreover, this will increase the survival rate as most of the eggs will be eaten by other marine animals or bigger fishes. This will ensure their species would not be extinct. The same theory to increase the chance of survival of offspring is also applied in all other animals, such as dogs, cats and turtles. Therefore, human is unique by limiting their offspring.

Human cognition

Michael Tomasello brought up the idea of 'ratchet effect' in human cultural evolution. "The most distinctive characteristic of human cultural evolution as a process is the way that modifications to an artefact or a social practice made by one individual or group of individuals often spread within the group, and then stay in place until some future individuals make further modifications --- and these then stay in place until still further modifications are made." The major part of the ratchet effect takes place in childhood.

Taboos

For Vezo people, what makes human different from animals is taboos. They believe that it is taboos which make us humans. Human beings practice a whole range of taboos which animals don't. Lambek (1992) noted that "taboos are distinctive at every level of social inclusion, from humanity viewed as a whole down to the individual." The Vezos have different taboos to show respect to animals. According to Astuti (2000), Vezos do not hunt or eat dolphins because they are generous towards human. Therefore, taboos on eating or hunting dolphins as dolphins saved people's lives by gently keeping them afloat and pushing them back to the land after their canoes had capsized.

However, children are too young to know about taboos and moral issues. They do not have taboos. If taboo is the unique quality of human beings, are children considered as animals? For me, children are considered as 'premature humans', which they do not have much experience on what is happening in the world. Even as I mentioned above, the brains undergoes human evolution and become more consciousness and awareness in their first four years of life. They are unwise and do not understand anything. However, as they grew up, and their perspectives will be changing. For example, after they have faced the life and death of their animal pets or toys, they will understand more about human death. And as what Astuti mentioned, 'this will contribute to transform their appreciation of the boundary between animals and people" Until they understand and appreciate the boundary between animals, they will become a 'mature human'. It takes 5-12 years for a child to become a 'mature human'. Therefore, taboos may be a unique quality for Homo sapiens.

AUTISM AND ANIMALS

"There is no cure," devastating words for any parent. Now add "it's your fault" and any parent would be shocked, hurt and even desperate. For many parents, especially mothers, that was the message for more than 30 years (Bettelheim). Told they were cold and had failed to bond with their children, many mothers had nowhere to turn. In the early days of autism there were few treatments available and only the inkling of understanding. How helpless it must have all felt.

It was during the 1960's that our modern understanding of autism began to take shape, and the controversy over treatments and the causes of autism took off. By the 1970's the origins of autism were beginning to become clear; the role of genetics helped to relieve some of the stigma that had surrounded a diagnosis of autism. As more and more children are diagnosed, boys far more than girls, the need for answers and effective treatments only becomes more urgent. There are a variety of treatments out there today touted as cures and overnight successes, none of which are backed by scientific proof, but there is growing research that animals can play a key part in managing the symptoms. Despite the fact that there is no cure for autism; animal therapy is an effective treatment for children with autism.

The discovery of autism has it roots in the study schizophrenia. In 1911, Eugene Bleuler who came up with the term schizophrenia noticed a segment of that population which was obviously different, and began to take note. Bleuler focused on a common symptom in this group, a disconnect with the outside world, and used the Greek word "autos" meaning "self" to coin the term "autism" ("autism"). Often considered retarded and psychologically flawed the outlook for these patients was grim. In the late 1930's, Hans Asperger was doing his own research in Vienna which was instrumental in defining this disorder. Eventually, around 1980, his research was used to determine a distinct diagnosis within the autism spectrum, what we now call Asperger's Syndrome (Lyons and Fitzgerald 2022). In 1943 Leo Kanner, another psychologist, was studying children here in the United States and changed things.

Kanner is considered a pioneer in the study of autism in children, and made great strides in diagnosis and treatment. The behaviors which set this group apart from all others, the severe delay or lack of verbal language, repetitive movements, an aversion to physical touch, and an obsessive restricted focus, all of which appear before age 3, became instrumental in diagnosing patients. During this time the common misconception that a failure to bond by mothers was the cause of these symptoms continued, it was also still often classified as a schizophrenic disorder (Rapin 302). Despite some major misconceptions about the origins of autism, there were great advances, one being that autism did not mean retardation, and as the study of autism continued, and so too the treatments.

In the early days treatments of autism was a grim prospect. Initially treated as a psychological disorder the treatments mirrored those for people labeled insane. Shock therapy was not uncommon, and is still used today, nor was isolation. It was a common practice to institutionalize children that were diagnosed. There they were often given drugs to subdue them, and "treat" their disorders. The most chronically used drugs, or first generation drugs, Thorazine and Haldol, were often used in conjunction with shock therapy from the 1950's until the 1970's (Watchel, Hermida and Dhossche). All of these treatments we now know to have side effects. The mild ones are hallucinations, prolonged contraction of muscles and medication induced Parkinsonism. If those are not enough to cope with, the severe ones like brain damage and involuntary muscle control can be permanent. These are the type of treatments that are responsible for the images of "zombie" patients wandering the halls of mental institutions. For parents coping with autism, these options weren't easy choices.

A few were able to overcome the current ideology, and try things that weren't common at the time. Temple Grandin, diagnosed in 1950, was a child that had one such parent. Her mother, Eustacia Cutler, went against all medical advice and treated her daughter with what at the time were unconventional and unproven methods (Cutler). Intense work on speech, social skills, public education (despite being labeled), and finally a college professor that recognized her abilities aside from her limitations has led her to be a well respected Professor of Animal Science at Colorado State University. Grandin, in her book Emergence: Labeled Autisic, says "I have read enough to know that there are still many parents, and yes, professionals too, who believe that 'once autistic, always autistic.' This dictum has meant sad and sorry lives for many children diagnosed, as I was in early life, as autistic. To these people, it is incomprehensible that the characteristics of autism can be modified and controlled. However, I feel strongly that I am living proof that they can" (16). Since that time, and indeed with her help, we have come a long way how we view and treat autism.

The treatments and diagnosis of autism have changed throughout the years just as much as the understanding of it has. "Autism is a developmental disorder, a defect in the systems which process incoming sensory information causes the child to over-react to some stimuli and under-react to others" (Grandin and Scariano). In simple terms this sensory issue is what causes the behaviors that have become the hallmark of autism. It is still difficult to measure the capabilities of a child with severely impaired communication skills, a lack of social awareness, and an inability to connect with the outside world. Some of the most successful treatments are physical therapy, occupational therapy, speech and language therapy, sensory integration therapy, and social skills therapy.

Many people are familiar with Physical Therapy (PT), which addresses things such as range of motion, balance, strength and dexterity. There are physical aspects of autism such as low muscle tone, toe walking and difficulty with coordination which can be addressed in PT. Treatment of these can also have other benefits as well. Things like attention, focus, and eye contact often improve during treatment (Spokes). It also has sets the stage for progressing into self care and self sufficient behaviors.

Occupational Therapy (OT) is less well known than PT, but is often more appropriate for autism patients. Unlike PT which focuses only on the mechanics of the body, OT takes it a step further and addresses daily tasks relevant to the patient. In a child that might include dressing, grooming, eating, and standing in line at school. All of which can be upsetting and even painful for an autistic child. It is also difficult for parents who are trying to go about daily life with an autistic child. Just getting bathed and dressed in the morning can turn into hour's long tests of patience and will, all before you even step out the door. OT is a great way to get individualized treatment for the varied issues autism can have. Some kids are just fine with tight socks and jeans, others will roll on the floor screaming in agony; knowing what can trigger the floor rolling and screaming, be it socks or sweatpants can be the difference between being homebound, and making it to the car in the morning. Quite often parents deal with these "meltdowns" which are a result of overstimulation. Just as in PT, OT also has other benefits such as structured time and repetitive activities, both shown to help alleviate some of the anxiety that autistic children suffer from (Spokes). It can also begin to address some of the sensory discomfort associated with autism, known as Sensory Integration Disorder.

Sensory Integration Therapy, or SIT, is a therapy more specific to autism. It is actually a combination of therapies used to address the sensory aspects of autism, like touch, smell, sound and taste, as well as the social ramifications. The major issue in autism is an imbalance in sensory reception (Bender). What may feel soft and comforting to us, can feel like needles to an autistic child. A routine trip to the grocery store can be simply awful for a child that can hear the piercing sound of the fluorescent lights. On top of that, imagine something as simple as your socks feeling like they are stabbing your ankles. These are the types of things the parent of an autistic child has to deal with on a daily basis, even a minute by minute basis. SIT is critical to functioning in our everyday world. Since autism is characterized by an inward focus, and disconnect from the outside world, it can be a challenge to coax such a child into interacting, even with the help of all of these therapies. This is when something almost magical, yet scientifically sound, can bridge that gap between a child stuck inside their own mind and the living world around them, the answer is animals.

There are several key factors that make animals a perfect fit for treatment of autistic children, our genetic predisposition toward animals domestic and otherwise (biophilia), the biological health benefits, and the practical benefits of a service animal; it is also the connection between animals and humans and the feedback loop that benefits us both (Tobin, et. al). There are many possible reasons for the effects of animals on autism, none of which have been proven in long term scientific studies. However, like the early days of treatments such as occupational therapy or sensory integration therapy that does not mean it doesn't work. As you will see there is scientific evidence of the positive effects of animals involved in the treatment and assessment of autism.

The first challenge in treating a disorder like autism is evaluating the severity. You need to know where the starting point is, and if you can't get a child to even look at a test what do you do? You use animals. It may seem trivial, but remember that autistic children find it extremely difficult, if not impossible, to make eye contact. It has been proven that typical testing, which often uses cards depicting human facial expressions, do not get a response from autistic children. This affects not only their ability to notice the surrounding environment, but also to read the faces of those around them, causing significant social dysfunction.

Just as with treatments, animals can play a pivotal role in evaluation and in measuring the progress of an autistic child. One of our first connections with animals was by observation in the time of the caveman this disposition towards animals is beneficial today as well. Simply replacing those faces with animal ones, there is a marked response (Hu). This is where biophilia and autism have common ground, the predisposition towards the natural world is not hampered by autism, it is not a social construct, it simply is. As simple as this may sound the use of animal representations in testing is still not widely accepted, utilized or understood. It is an emerging topic of study (New). If pictures of animals elicit a response, what kind of reaction do living ones have?

The most widely acknowledge aspect of autism is the lack of verbal communication, which causes severe social deficits. Here is where animals, especially domesticated, appear almost magical. Unlike people who have a myriad of potential responses, both physical and verbal, animal communication is more predictable. A set pattern of behavior and response allows autistic children communicate without the clutter human interaction can cause, thus avoiding overstimulation. There is some research which uses robots to mimic this set behavior and response pattern in conjunction with autism treatments, but it lacks the feedback that animals provide (Jarriot). Animals being largely nonverbal creatures; can read our intentions in the subtle shift of our bodies, the look in our eyes, and the smell of our skin (McCullough). The pressure to communicate, which can be overwhelming in autism, is moot. This nonverbal communication makes animals a perfect choice for autism treatment. This does not mean that therapy is not needed, in fact it is the opposite; animals and therapy go hand in hand.

The therapies commonly used today to treat the symptoms of autism can be difficult at best. It is a deliberate exposure to the very things that can irritate an autistic child; however the addition of an animal in therapy sessions has been shown to improve social interactions and language, the top two issues those with autism face. Using animals for treatment in this manner is called a biocentric approach because it goes hand in hand with the theory of biophilia.

Therapy can present a challenge for an autistic child; verbal directions, physical contact, and the sheer overwhelming aspect of the environment can make it nearly impossible to function. A study titled "Occupational Therapy Incorporating Animals" profiled the use of animals during therapy sessions and "found that children with autism demonstrated fewer stereotyped behaviors such as hand-posturing and humming, and increased socially appropriate behavior such as joining the therapist in simple games, when a friendly therapy dog was introduced into the sessions" (pg. 270). We may not know exactly why this works, but we do know that "incorporating animals in the occupational therapy session was unique in that it allowed the children to learn to interpret and respond to the less complex social and behavioral cues of animals, which may provide a bridge to learning to interpret the more subtle behavior of human beings" (Sams). Animals of all kinds are not only beneficial in conjunction with therapy, but as a therapy themselves.

Hippotherapy, also known as equine therapy, is the use of the horses as treatment, and encompasses physical, occupational and speech therapy. Each person is assessed and a treatment plan is made, just as in occupational therapy individual needs are the focus, which is especially useful in autism since no two autistic children are alike. Horses have a similar pelvic movement as humans when walking; feeling this sensory input, which is steady and rhythmic, deals with the sensory input imbalance intrinsic to autism (Bender). It also encompasses multiple sensory areas like smell, vision, touch, movement, balance and sound. Learning to process this sensory input can lead to better functioning in environments where overstimulation is rife, and decrease the "meltdowns." Aside from the physical benefits of horse therapy there are psychological ones as well. A recent study in the journal, Occupational Therapy in Mental Health, noted that "violition", or motivation, is adversely affected by the impairments of autism. The study was able to show an increase in violition, or participation, and an increase in interest in new activities as a direct result of Hippotherapy (Taylor, et al.). The effects of animal therapy, although not always explained, are both obvious and often dramatic.

The benefits of animals involved in therapy are not limited to physical and occupational therapy. There are notable physical effects as well. Science has known oxytocin for some time, but its true functions are just beginning to be revealed. It is now known that increased levels will reduce high blood pressure, lower heart rate and decrease stress responses (Olmert). Where can we get this? Animals. The simple act of petting an animal creates a feedback loop in which oxytocin levels are increased in both the person and the animal. For a parent much of our time can be spent calming our autistic child, which is especially difficult after a "meltdown." It can take hours for a child to return to a state calm enough to tolerate physical touch. A pet, cat or dog, can reduce the time it takes to recover from, and even sometimes avoid, these meltdowns. Animals often have an instant impact on autistic children, and the benefits last long past the immediate. Even though no one single treatment works for all, it is most often that animals create that immediate positive response which can last long past the therapy session. Research "has indicated that the gains associated with the involvement of the animals were valuable and observable in the children's daily lives" (Sams, Fortney, and Willenbring 273). The simple presence of an animal created immediate and lasting results, which then begs the question what would a trained animal do for autism? Service animals, although currently controversial, are another way animals can improve the daily life of someone with autism.

Service animals are a common treatment for disabled individuals. We are familiar with ones used to assist the physically handicapped. In recent years the use of service animals for children with autism have become a more recognized and accepted option. A service animal is an ideal choice for the physical aspects of a disability, but more now than ever we are seeing the positive effects on "invisible disabilities" such as autism (Skloot). Though we most often relate service animal with dogs, there are many types of animals which have been proven to benefit autistic children, even guinea pigs. An animal can be trained to do the more obvious tasks of alerting parents when an autistic child wanders away, to the less obvious ones of sensing increased agitation and responding.

Not only are there benefits that arise from the trained aspects of a service animal, there are coincidental benefits which can be an immense help in the area of social skills. A service animal can be a conversational cue for autistic children who have poor social and language skills. An animal trained to read hand signals can be used when a child lacks language, and the training of hand cues itself benefits the child by teaching the verbal counterparts. Also by removing the stress of verbal communication the service animal becomes an oasis, which can reduce meltdowns. There are many ways a service animal benefits an autistic child, but they are not appropriate for everyone. As with any treatment, not all autistic children respond the same. The severity of impairment and temperament also factor into whether a service animal is appropriate. Acceptance of service animals can also be an issue.

Service animals are under fire at all levels. Simple getting into everyday places like a grocery store, restaurant, bus or even the doctor can be a hassle. The laws regulating service animals are undergoing change right now, and legal cases are cropping up across the nation. In Oregon, an autistic boy has been denied his service dog while in school (Owen). Since trained service dogs for autism, especially in school, are still rare the outcome is not clear. Autism is getting more attention and more study is needed, but the need for effective treatments will remain as long as autism remains.

Although there is ongoing research and development in the area of animals and autism treatment, it is still considered a "new" field of study. Doubt among scientist remains to the origins of autism, and hence even more so over the treatments. Many people hesitate to ascribe to any specific therapy; however, as during the time of Temple Grandin when therapies which are common today were either not available or not recognized as treatment, that does not lessen their effectiveness. That did not deter her mother, and it should not deter us today. For Temple animals were not only a fascination, but a gateway for her to connect to the world around her, and in turn to the rest of us. We should not shun any treatment that shows not only promise, but results.

It is a fact, there is still no cure for autism, but there is hope, where none was before. As we understand the origins of this disorder, so too do we understand the treatments. It has been shown time and again our undeniable connection to animals. The title of Temple Grandin's latest book is "Animals Make us Human," and I think she may be right. Animals have a natural ability to calm and comfort us, and as the evidence can attest to heal us as well.

How to Teach a Dog to Stay

Training a dog involves the positive strengths and encouragement. To train a dog to stay when commanded is a useful advantage in many situations. It requires proper time, patience and determination though it is a simple process. This type of behavior of the dog is beneficial to the dog's owner. When the dog gets over excited or does not behave properly, then the owner can settle his dog by commanding him to stay. Effective training can be provided to the dog in a calm and quiet location so that the dog can concentrate on the owner's training without any distraction. Do not try to train your dog when he is in the excited mood.

The first thing to start with the training is to make your dog sit and get its attention towards you. Do not provide any reward to the dog before the training. Once the dog is seated in front of you, just say the word "stay". After sometime, move to the side of the dog and behind the dog but do not move away from your dog. If your dog does not move then offer him a treat. You can try to go a bit away from the dog. But, he gets up from his place and moves around you, then place him in the original location and command him to sit. Repeat this process until the dog gets seated in his position even if you move away from him. In the initial stage, let your dog sit only for 15 to 20 minutes.

Once the dog continues to sit and stay at the same place even if you are moving, then proceed further to the next step. Now start increasing the distance from the dog. Instruct the dog to sit at the same place as before and you try to move some yards away from your dog. Reward the dog as before if he does not move from its position. Start increasing the time as the dog understands the command to stay. Train your dog in such a way that he will remain in the seated position without a move when you are out of that room for some time.

Here are some additional tips with basic techniques to successfully train a dog to stay. During the training session of a dog, you should remain generous in treating your dog as well as in giving the rewards. Try to keep the sessions of small duration at least in the beginning. You can take several sessions in a day of 10-15 minutes. Instruct your dog by making it as a game. The dog cannot immediately understand your commands. So, if it does not perform up to your mark, then instead of punishing him, just do not reward him. Remember that not to provide any kind of training when the dog is in dangerous situation or any bad circumstances. Try to get him out of that situation and make him feel relax by diverting his mind in playing his favorite games.

ANIMAL INTELLIGENCE AND EVOLUTION OF THE HUMAN MIND

The human brain lacks conspicuous characteristics-such as relative or absolute size-that might account for humans' superior intellect.

Researchers have found some clues to humanity's aptitude on a smaller scale, such as more neurons in our brain's outermost layer.

Human intelligence may be best likened to an upgrade of the cognitive capacities of nonhuman primates rather than an exceptionally advanced form of cognition.

Subtle refinements in brain architecture, rather than large-scale alterations, make us smarter than other animals.

As far as we know, no dog can compose music, no dolphin can speak in rhymes, and no parrot can solve equations with two unknowns. Only humans can perform such intellectual feats, presumably because we are smarter than all other animal species-at least by our own definition of intelligence.

Of course, intelligence must emerge from the workings of the three-pound mass of wetware packed inside our skulls. Thus, researchers have tried to identify unique features of the human brain that could account for our superior intellectual abilities. But, anatomically, the human brain is very similar to that of other primates because humans and chimpanzees share an ancestor that walked the earth less than seven million years ago.

Accordingly, the human brain contains no highly conspicuous characteristics that might account for the species' cleverness. For instance, scientists have failed to find a correlation between absolute or relative brain size and acumen among humans and other animal species. Neither have they been able to discern a parallel between wits and the size or existence of specific regions of the brain, excepting perhaps Broca's area, which governs speech in people. The lack of an obvious structural correlate to human intellect jibes with the idea that our intelligence may not be wholly unique: studies are revealing that chimps, among various other species, possess a diversity of humanlike social and cognitive skills.

Nevertheless, researchers have found some microscopic clues to humanity's aptitude. We have more neurons in our brain's cerebral cortex (its outermost layer) than other mammals do. The insulation around nerves in the human brain is also thicker than that of other species, enabling the nerves to conduct signals more rapidly. Such biological subtleties, along with behavioral ones, suggest that human intelligence is best likened to an upgrade of the cognitive capacities of nonhuman primates rather than an exceptionally advanced form of cognition.

Smart Species

Because animals cannot read or speak, their aptitude is difficult to discern, much less measure. Thus, comparative psychologists have invented behavior-based tests to assess birds' and mammals' abilities to learn and remember, to comprehend numbers and to solve practical problems. Animals of various stripes-but especially nonhuman primates-often earn high marks on such action-oriented IQ tests. During World War I, German psychologist Wolfgang Köhler, for example, showed that chimpanzees, when confronted with fruit hanging from a high ceiling, devised an ingenious way to get it: they stacked boxes to stand on to reach the fruit. They also constructed long sticks to reach food outside their enclosure. Researchers now know that great apes have a sophisticated understanding of tool use and construction.

Psychologists have used such behavioral tests to illuminate similar cognitive feats in other mammals as well as in birds. Pigeons can discriminate between male and female faces and among paintings by different artists; they can also group pictures into categories such as trees, selecting those belonging to a category by pecking with their beaks, an action that often brings a food reward. Crows have intellectual capacities that are overturning conventional wisdom about the brain.

Behavioral ecologists, on the other hand, prefer to judge animals on their street smarts-that is, their ability to solve problems relevant to survival in their natural habitats-rather than on their test-taking talents. In this view, intelligence is a cluster of capabilities that evolved in response to particular environments. Some scientists have further proposed that mental or behavioral flexibility, the ability to come up with novel solutions to problems, is another good measure of animal intellect. Among birds, green herons occasionally throw an object in the water to lure curious fish-a trick that, ornithologists have observed, has been reinvented by groups of these animals living in distant locales. Even fish display remarkable practical intelligence, such as the use of tools, in the wild. Cichlid fish, for instance, use leaves as "baby carriages" for their egg masses.

Animals also can display humanlike social intelligence. Monkeys engage in deception, for example; dolphins have been known to care for another injured pod member (displaying empathy), and a whale or porpoise may recognize itself in the mirror. Even some fish exhibit subtle kinds of social skills. Behavioral ecologist Redouan Bshary of the University of Neuchâtel in Switzerland and his colleagues described one such case in a 2006 paper. Bony fish such as the so-called cleaner wrasse (Labroides dimidiatus) cooperate and remove parasites from the skin of other fish or feed on their mucus. Bshary's team found that bystander fish spent more time next to cleaners the bystanders had observed being cooperative than to other fish. Humans, the authors note, tend to notice altruistic behavior and are more willing to help do-gooders whom they have observed doing favors for others. Similarly, cleaner wrasses observe and evaluate the behavior of other finned ocean denizens and are more willing to help fish that they have seen assisting third parties.

From such studies, scientists have constructed evolutionary hierarchies of intelligence. Primates and cetaceans (whales, dolphins and porpoises) are considered the smartest mammals. Among primates, humans and apes are considered cleverer than monkeys, and monkeys more so than prosimians. Of the apes, chimpanzees and bonobos rank above gibbons, orangutans and gorillas. Dolphins and sperm whales are supposedly smarter than nonpredatory baleen whales such as blue whales. Among birds, scientists consider parrots, owls and corvids (crows and ravens) the brightest. Such a pecking order argues against the idea that intelligence evolved along a single path, culminating in human acumen. Instead intellect seems to have emerged independently in birds and mammals and also in cetaceans and primates.

Heavy Thoughts?

What about the brain might underlie these parallel paths to astuteness? One candidate is absolute brain size. Although many studies have linked brain mass with variations in human intelligence [see "High-ÂAptitude Minds," by Christian Hoppe and Jelena Stojanovic], size does not always correlate with smarts in different species. For example, clever small animals such as parrots, ravens, rats and relatively diminutive apes have brains of modest proportions, whereas some large animals such as horses and cows with large brains are comparatively dim-witted. Brain bulk cannot account for human intelligence either: At eight to nine kilograms, sperm and killer whale brains far outweigh the 1.4 kilograms of neural tissue inside our heads. As heavy as five kilograms, elephant brains are also much chunkier than ours.

Relative brain size-the ratio of brain to body mass-does not provide a satisfying explanation for interspecies differences in smarts either. Humans do compare favorably with many medium and large species: our brain makes up approximately 2 percent of our body weight, whereas the blue whale's brain, for instance, is less than one 100th of a percent of its weight. But some tiny, not terribly bright animals such as shrews and squirrels win out in this measure. In general, small animals boast relatively large brains, and large animals harbor relatively small ones. Although absolute brain mass increases with body weight, brain mass as a proportion of body mass tends to decrease with rising body weight.

Another cerebral yardstick that scientists have tried to tie to intelligence is the degree of encephalization, measured by the encephalization quotient (EQ). The EQ expresses the extent to which a species' relative brain weight deviates from the average in its animal class, say, mammal, bird or amphibian. Here the human brain tops the list: it is seven to eight times larger than would be expected for a mammal of its weight. But EQ does not parallel intellect perfectly either: gibbons and some capuchin monkeys have higher EQs than the more intelligent chimpanzees do, and even a few proÂsimÂians-the earliest evolved primates alive today-have higher EQs than gorillas do.

Or perhaps the size of the brain's outermost layer, the cerebral cortex-the seat of many of our cognitive capacities-is the key. But it turns out that the dimensions of the cerebral cortex depend on those of the entire brain and that the size of the cortex constitutes no better arbiter of a superior mind. The same is true for the prefrontal cortex, the hub of reason and action planning. Although some brain researchers have claimed in the past that the human prefrontal cortex is exceptionally large, recent studies have shown that it is not. The size of this structure in huÂmans is comparable to its size in other Âprimates and may even be relatively small as compared with its counterpart in elephants and cetaceans.

The lack of a large-scale measure of the human brain that could explain our performance may reflect the idea that human intellect may not be totally inimitable. Apes, after all, understand cause and effect, make and use tools, produce and comprehend language, and lie to and imitate others. These primates may even possess a theory of mind-the ability to understand another animal's mental state and use it to guide their own behavior. Whales, dolphins and even some birds boast some of these mental talents as well. Thus, adult humans may simply be more intuitive and facile with tools and language than other species are, as opposed to possessing unique cognitive skills.

Networking

Fittingly, researchers have found the best correlates for intelligence by looking at a much smaller scale. Brains consist of nerve cells, or neurons, and supporting cells called glia. The more neurons, the more extensive and more productive the neuronal networks can be-and those networks determine varied brain functions, including perception, memory, planning and thinking. Large brains do not automatically have more neurons; in fact, neuronal density generally decreases with increasing brain size because of the additional glial cells and blood vessels needed to support a big brain.

Humans have 11.5 billion cortical neurons-more than any other mammal, because of the human brain's high neuronal density. Humans have only about half a billion more cortical neurons than whales and elephants do, however-not enough to account for the significant cognitive differences between humans and these species. In addition, however, a brain's information-processing capacity depends on how fast its nerves conduct electrical impulses. The most rapidly conducting nerves are swathed in sheaths of insulation called myelin. The thicker a nerve's myelin sheath, the faster the neural impulses travel along that nerve. The myelinated nerves in the brains of whales and elephants are demonstrably thinner than they are in primates, suggesting that information travels faster in the human brain than it does in the brains of nonprimates.

What is more, neuronal messages must travel longer distances in the relatively large brains of elephants and whales than they do in the more compact human brain. The resulting boost in information-processing speed may at least partly explain the disparity in aptitude between humans and other big-brained creatures.

Among humans' cerebral advantages, language may be the most obvious. Various animals can convey complex messages to other members of their species; they can communicate about objects that are not in sight and relay information about individuals and events. Chimpanzees, gorillas, dolphins and parrots can even understand and use human speech, gestures or symbols in constructions of up to about three words. But even after years of training, none of these creatures develops verbal skills more advanced than those of a three-year-old child.

In humans, grammar and vocabulary all but explode at age three. This timing corresponds with the development of Broca's speech area in the left frontal lobe, which may be unique to humans. That is, scientists are unsure whether a direct precursor to this speech region exists in the nonhuman primate brain. The absence of an intricately wired language region in the brains of other species may explain why, of all animals, humans alone have a language that contains complex grammar. Researchers date the development of human grammar and syntax to between 80,000 and 100,000 years ago, which makes it a relatively recent evolutionary advance. It was also one that probably greatly enhanced human intellect.

COEVOLUTION

In biology, coevolution is "the change of a biological object triggered by the change of a related object.". In other words, when changes in at least two species' genetic compositions reciprocally affect each other's evolution, coevolution has occurred.

There is evidence for coevolution at the level of populations and species. For example, the concept of coevolution was briefly described by Charles Darwin in *On the Origin of Species*, and developed in detail in *Fertilisation of Orchids*. It is likely that viruses and their hosts may have coevolved in various scenarios.

However, there is little evidence of coevolution driving large-scale changes in Earth's history, since abiotic factors such as mass extinction and expansion intoecospace seem to guide the shifts in the abundance of major groups. One specific example is the evolution of high-crowned teeth in grazers when grasslands spread through North America. Long held up as an example of coevolution, we now know that these events happened independently.

Coevolution can occur at many biological levels: it can be as microscopic as correlated mutations between amino acids in a protein, or as macroscopic as covarying traits between different species in an environment. Each party in a coevolutionary relationship exerts selective pressures on the other, thereby affecting each other's evolution. Coevolution of different species includes the evolution of a host species and its parasites (host–parasite coevolution), and examples of mutualism evolving through time. Evolution in response to abiotic factors, such as climate change, is not biological coevolution (since climate is not alive and does not undergo biological evolution).

The general conclusion is that coevolution may be responsible for much of the genetic diversity seen in normal populations including: blood plasma polymorphism, protein polymorphism, histocompatibility systems, etc.

The parasite host relationship is probably what drove the prevalence of sexual reproduction over the more efficient asexual reproduction. It seems that most sources determine that when a host is infected by a parasite, sexual reproduction affords a chance of resistance, through variation in the next generation, giving sexual reproduction viability for fitness not seen in the asexual reproduction, which would only generate another generation of the organism susceptible to infection by the same parasite.

Coevolution is primarily a biological concept, but has been applied by analogy to fields such as computer science, sociology / International Political Economy and astronomy.

Models

One model of coevolution was Leigh Van Valen's Red Queen's Hypothesis, which states that "for an evolutionary system, continuing development is needed just in order to maintain its fitness relative to the systems it is co-evolving with". This hypothesis predicts that sexual reproduction allows a host to stay just ahead of its parasite by a generation, similar to the Red Queen in "Through the Looking Glass". ...always running ..... Just ahead. The essence is that the host reproduces sexually giving it immunity over its parasite, which then evolves in response. This requires the next generation to repeat the sequence. Emphasizing the importance of sexual conflict, Thierry Lodé described the role of *antagonist interactions* in evolution, giving rise to a concept of *antagonist coevolution*. Coevolution branching strategies for asexual population dynamics in limited resource environments have been modeled using the generalized Lotka–Volterra equations. A model based on adaptive dynamics and experimental data of floral and proboscis lengths, as well as nectar consumed and pollen deposited during the pollination of the long-tubed iris (*Lapeirousia anceps*) by the long-proboscid fly (*Moegistorhynchus longirostris*) has generated diverse coevolutionary dynamics, including two types of Red Queen dynamics, evolutionary branching (backed by observations of coexisting irises of short and long tubes in a single population) and trap.

Coevolution types

Coevolution can occur between pairs of entities (often referred to as pairwise coevolution) exists, with examples includingpredator and prey, host and symbiont or host and parasite. However, some instances of coevolution are less clearcut: a species may evolve in response to more than one species.

If this coevolution is occurring in a non-additive way, then this type of coevolution is known as diffuse coevolution.

A general characterization that can be made of many viruses, widely known to be obligate parasites, is that they coevolved alongside their respective hosts. This is suggested by the similar genetic arrangement between virus and host.Correlated mutations between the two species enter them into an evolution arms race: the host must develop a defense mechanism to overcome the parasite, and the parasite must overcome the new defense mechanism in order to persist and reproduce. Whichever organism, host or parasite, that cannot keep up with the other will be eliminated from their habitat, as the species with the higher average population fitness survives. This race is known as the Red Queen hypothesis.

Escape and radiate coevolution occurs when an organism under selectional constraints by predation develops a defense mechanism allowing it to "escape", and then "radiate" into another species entirely. The predator then evolves accordingly, and the process continues.

Coevolution in the fossil record

Many examples of coevolution have been observed among living species but coevolution has not been conclusively shown in the fossil record. This may be because the fossil record does not tend to preserve high resolution data (e.g. on the level of species) and because coevolution does not drive large-scale changes in Earth's history. Even classic examples of coevolution in the fossil record such as the interaction between bees and the flowers they pollinate or the development of grazers in spreading grasslands are little supported by evidence. In the case of bees and flowers, these two organisms evolved at different times - Archaefructus is a flowering plant from 125 million years ago but bees are known from nests 220 million years ago. Similarly in the case of grazing ungulates and the spread of grasslands - the timing in evolution of the two organisms is very different - this is called the Grit, not grass hypothesis.

Specific examples

Hummingbirds and ornithophilous flowers

Hummingbirds and ornithophilous (bird-pollinated) flowers have evolved a mutualistic relationship. The flowers have nectarsuited to the birds' diet, their color suits the birds' vision and their shape fits that of the birds' bills. The blooming times of the flowers have also been found to coincide with hummingbirds' breeding seasons. The floral characteristics of ornithophilous plants vary greatly among each other compared to closely related insect-pollinated species. These flowers also tend to be more ornate, complex, and showy than their insect pollinated counterparts. It is generally agreed upon that plants formed coevolutionary relationships with insects first, and ornithophilous species diverged at a later time. There is not much scientific support for instances of the reverse of this divergence: from ornithophily to insect pollination. The diversity in floral phenotype in ornithophilous species, and the relative consistency observed in bee-pollinated species can be attributed to the direction of the shift in pollinator preference.

Flowers have converged to take advantage of similar birds. Flowers compete for pollinators, and adaptations reduce unfavourable effects of this competition. The fact that birds can fly during inclement weather can make them more efficient pollinators in cases in which bees and other insects are inactive. Ornithophily may have arisen for this reason in isolated environments with poor insect colonization or areas with plants which flower in the winter. Bird-pollinated flowers usually have higher volumes of nectar and higher sugar production than those pollinated by insects. This meets the birds' high energy requirements, which are the most important determinants of their flower choice. In experiments with monkeyflowers, an increase in red pigment in petals and flower nectar volume has been shown to noticeably reduce the proportion of pollination by bees as opposed to solely attract hummingbirds; while greater flower surface area was shown to increase the amount of bee pollination. Additional red coloration of flowers significantly decreased bee visitation but seemingly had no effect on the frequency of hummingbird visitation. Thus, hummingbirds may not necessarily have an innate preference for red and the high concentration of these red pigments in the flowers of *M. cardinalis* could potentially function primarily to discourage bee visitation. In first generation hybrids of these two species, the composition of pollinator visitors (59% bees; 1,744 visits) was exactly in between the two parental species, implying a strong genetic component to pollinator coevolution. In similar experiments with *Penstemon* flowers, pollen transfer by both bees and hummingbirds was recorded with two closely related species species differing in main pollinator. The results suggest that the flower traits that discourage bee pollination may be even more influential on the flowers' evolutionary change than 'pro-bird' adaptations are. However, adaptation 'towards' birds and adaptation 'away' from bees can and do happen simultaneously. Following their respective breeding seasons, several species of hummingbirds occur at the same locations in North America, and several hummingbird flowers bloom simultaneously in these habitats. These flowers seem to have converged to a common morphology and color. Different lengths and curvatures of the corolla tubes can affect the efficiency of extraction in hummingbird species in relation to differences in bill morphology. Tubular flowers force a bird to orient its bill in a particular way when probing the flower, especially when the bill and corolla are both curved; this also allows the plant to place pollen on a certain part of the bird's body. This opens the door for a variety of morphological co-adaptations.

An important requirement for attraction is conspicuousness to birds, which reflects the properties of avian vision and habitat features. Birds have their greatest spectral sensitivity and finest hue discrimination at the red end of the visual spectrum, so red is particularly conspicuous to them. Hummingbirds may also be able to see ultraviolet "colors". The prevalence of ultraviolet patterns and nectar guides in nectar-poor entomophilous (insect-pollinated) flowers warns the bird to avoid these flowers.

Hummingbirds form the family Trochilidae, whose two subfamilies are the Phaethornithinae (hermits) and the Trochilinae. Each subfamily has evolved in conjunction with a particular set of flowers. Most Phaethornithinae species are associated with large monocotyledonous herbs, while the Trochilinae prefer dicotyledonous plant species.

Angraecoid orchids and African moths

Angraecoid orchids and African moths coevolve because the moths are dependent on the flowers for nectar and the flowers are dependent on the moths to spread pollen so they can reproduce. Coevolution has led to deep flowers and moths with long probosci.

Old world swallowtail and fringed rue

An example of antagonistic coevolution is the old world swallowtail (*Papilio machaon*) caterpillar living on the fringed rue (*Ruta chalepensis*) plant. The rue produces etheric oils which repel plant-eating insects. The old world swallowtail caterpillar developed resistance to these poisonous substances, thus reducing competition with other plant-eating insects.

Garter snake and rough-skinned newt

Coevolution of predator and prey species is illustrated by the Rough-skinned newt (*Taricha granulosa*) and the common garter snake (*Thamnophis sirtalis*). The newts produce a potent neurotoxin that concentrates in their skin. Garter snakes have evolved resistance to this toxin through a series of geneticmutations, and prey upon the newts. The relationship between these animals has resulted in an evolutionary arms race that has driven toxin levels in the newt to extreme levels. This is an example of coevolution because differential survival caused each organism to change in response to changes in the other.

California buckeye and pollinators

When beehives are populated with bee species that have not coevolved with the California buckeye (*Aesculus californica*), sensitivity to aesculin, a neurotoxin present in its nectar, may be noticed; this sensitivity is only thought to be present in honeybees and other insects that did not coevolve with *A. californica*.

Acacia ant and bullhorn acacia tree

The acacia ant (*Pseudomyrmex ferruginea*) protects the bullhorn acacia (*Acacia cornigera*) from preying insects and from other plants competing for sunlight, and the tree provides nourishment and shelter for the ant and its larvae. Nevertheless, some ant species can exploit trees without reciprocating, and hence have been given various names such as 'cheaters', 'exploiters', 'robbers' and 'freeloaders'. Although cheater ants do important damage to the reproductive organs of trees, their net effect on host fitness is difficult to forecast and not necessarily negative.

7

Biogeography

Biogeography is the study of the distribution of species and ecosystemsin geographic space and through geological time. Organisms and biological communities often vary in a regular fashion along geographic gradients of latitude, elevation, isolation and habitat area. Phytogeography is the branch of biogeography that studies the distribution of plants. Zoo geography is the branch that studies distribution of animals.

Knowledge of spatial variation in the numbers and types of organisms is as vital to us today as it was to our early human ancestors, as we adapt to heterogeneous but geographically predictable environments. Biogeography is an integrative field of inquiry that unites concepts and information from ecology, evolutionary biology, geology, and physical geography.

Modern biogeographic research combines information and ideas from many fields, from the physiological and ecological constraints on organismaldispersal to geological and climatological phenomena operating at global spatial scales and evolutionary time frames.

The short-term interactions within a habitat and species of organisms describe the ecological application of biogeography. Historical biogeography describes the long-term, evolutionary periods of time for broader classifications of organisms. Early scientists, beginning with Carl Linnaeus, contributed theories to the contributions of the development of biogeography as a science. Beginning in the mid-18th century, Europeans explored the world and discovered the biodiversity of life. Linnaeus initiated the ways to classify organisms through his exploration of undiscovered territories.

The scientific theory of biogeography grows out of the work of Alexander von Humboldt (1769–1859), Hewett Cottrell Watson (1804–1881), Alphonse de Candolle (1806–1893), Alfred Russel Wallace (1823–1913), Philip Lutley Sclater (1829–1913) and other biologists and explorers.

INTRODUCTION

The patterns of species distribution across geographical areas can usually be explained through a combination of historical factors such as: speciation, extinction, continental drift, and glaciation. Through observing the geographic distribution of species, we can see associated variations in sea level, river routes, habitat, and river capture. Additionally, this science considers the geographic constraints of landmass areas and isolation, as well as the available ecosystem energy supplies.

Over periods of ecological changes, biogeography includes the study of plant and animal species in: their past and/or present living *refugium* habitat; their interim living sites; and/or their survival locales. As writer David Quammen put it, "...biogeography does more than ask *Which species?* and *Where*. It also asks *Why?* and, what is sometimes more crucial,*Why not?*."

Modern biogeography often employs the use of Geographic Information Systems (GIS), to understand the factors affecting organism distribution, and to predict future trends in organism distribution. Often mathematical models and GIS are employed to solve ecological problems that have a spatial aspect to them.

Biogeography is most keenly observed on the world's islands. These habitats are often much more manageable areas of study because they are more condensed than larger ecosystems on the mainland. Islands are also ideal locations because they allow scientists to look at habitats that new species have only recently colonized and can observe how they disperse throughout the island, the success they achieve in these places, and they can then apply this information to similar mainland habitats. Islands are very diverse in their biomes, ranging from the tropical to arctic climates. This diversity in habitat allows for a wider range of species study in different parts of the world.

One scientist who recognized the importance of these geographic locations was Charles Darwin, who remarked in his journal "The Zoology of Archipelagoes will be well worth examination". Two chapters in *On the Origin of Species* were devoted to geographical distribution.

History

18th Century

The first discoveries that contributed to the development of biogeography as a science began in the mid-18th century, as Europeans explored the world and discovered the biodiversity of life. During the 18th century most views on the world were shaped around religion and for many natural theologists, the bible. Carl Linnaeus, in the mid-18th century, initiated the ways to classify organisms through his exploration of undiscovered territories. When he noticed that species were not as perpetual as he believed, he developed the

Mountain Explanation to explain the distribution of biodiversity. When Noah's ark landed on Mount Ararat and the waters receded, the animals dispersed throughout different elevations on the mountain. This showed different species in different climates proving species were not constant. Linnaeus' findings set a basis for ecological biogeography. Through his strong beliefs in Christianity, he was inspired to classify the living world, which then gave way to additional accounts of secular views on geographical distribution. He argued that the structure of an animal was very closely related to its physical surroundings. This was important to a George Louis Buffon's rival theory of distribution.

Closely after Linnaeus, Georges-Louis Leclerc, Comte de Buffon observed shifts in climate and how species spread across the globe as a result. He was the first to see different groups of organisms in different regions of the world. Buffon saw similarities between some regions which led him to believe that at one point continents were connected and then did water separate them and cause differences in species. His hypotheses were described by his books, Histoire Naturelle, and Générale et Particulière, in which he argued that varying geographical regions would have different forms of life. This was inspired by his observations comparing the Old and New World, as he determined distinct variations of species from the two regions. Buffon believed there was a single species creation event, and that different regions of the world were homes for varying species, which is an alternate view than that of Linnaeus. Buffon's Law eventually became a principle of biogeography by explaining how similar environments were habitats for comparable types of organisms. Buffon also studied fossils which led him to believe that the earth was over tens of thousands of years old, and that humans had not lived there long in comparison to the age of the earth.

Following this period of exploration came the Age of Enlightenment in Europe, which attempted to explain the patterns of biodiversity observed by Buffon and Linnaeus. At the end of the 18th century, Alexander von Humboldt, known as the "founder of plant geography", developed the concept of physique generale to demonstrate the unity of science and how species fit together. As one of the first to contribute empirical data to the science of biogeography through his travel as an explorer, he observed differences in climate and vegetation. The earth was divided into regions which he defined as tropical, temperate, and arctic and within these regions there were similar forms of vegetation. This ultimately enabled him to create the isotherm, which allowed scientists to see patterns of life within different climates. He contributed his observations to findings of botanical geography by previous scientists, and sketched this description of both the biotic and abiotic features of the earth in his book, Cosmos.

Augustin de Candolle contributed to the field of biogeography as he observed species competition and the several differences that influenced the discovery of the diversity of life. He was a Swiss botanist and created the first Laws of Botanical Nomenclature in his work, Prodromus. He discussed plant distribution and his theories eventually had a great impact on Charles Darwin, who was inspired to consider species adaptations and evolution after learning about botanical geography. De Candolle was the first to describe the differences between the small-scale and large-scale distribution patterns of organisms around the globe.

19th Century

In the 19th century, several additional scientists contributed new theories to further develop the concept of biogeography. Charles Lyell, being one of the first contributors in the 19th century, developed the Theory of Uniformitarianism after studying fossils. This theory explained how the world was not created by one sole catastrophic event, but instead from numerous creation events and locations. Uniformitarianism also introduced the idea that the Earth was actually significantly older than was previously accepted. Using this knowledge, Lyell concluded that it was possible for species to go extinct. Since he noted that earth's climate changes, he realized that species distribution must also change accordingly. Lyell argued that climate changes complemented vegetation changes, thus connecting the environmental surroundings to varying species. This largely influenced Charles Darwin in his development of the theory of evolution.

Charles Darwin was a natural theologist who studied around the world, and most importantly in the Galapagos Islands. Darwin introduced the idea of evolution, as he theorized against previously accepted ideas that species were static or unchanging. His contributions to biogeography and the theory of evolution were different from those of other explorers of his time, because he developed a mechanism to describe the ways that species changed. His influential ideas include the development of theories regarding the struggle for existence and natural selection. Darwin's theories started a biological segment to biogeography and empirical studies, which enabled future scientists to develop ideas about the geographical distribution of organisms around the globe.

Later on in the 19th century, Alfred Russel Wallace became commonly known as the father of biogeography after his contributions regarding the distribution of organisms. He observed the butterfly and bird distributions in comparison to the presence or absence of geographical barriers. Additionally, he argued that the number of organisms present in a community was dependent on the amount of food resources in the particular habitat. Wallace participated in fieldwork researching the habits, breeding and migration tendencies, and feeding behavior of thousands of species. Wallace studied the distribution of flora and fauna in the Amazon Basin and the Malay Archipelago in the mid-19th century. Wallace and Sclater saw biogeography as a source of support for the theory of evolution. Key findings, such as the sharp difference in fauna either side of the Wallace Line, and the sharp difference that existed between North and South America prior to their relatively recent faunal interchange, can only be understood in this light. Using these factors, he studied their influence on the survival of species. Wallace believed species were dynamic by responding to biotic and abiotic factors.

20th Century

Moving on to the 20th century, Alfred Wegener introduced the Theory of Continental Drift in 1912, though it was not widely accepted until the 1960s. This theory was revolutionary because it changed the way that everyone thought about species and their distribution around the globe. The theory explained how continents were formerly joined together in one large landmass, Pangea, and slowly drifted apart due to the movement of the plates below Earth's surface. The evidence for this theory is in the geological similarities between varying locations around the globe, fossil comparisons from different continents, and the jigsaw puzzle shape of the landmasses on Earth. Though Wegener did not know the mechanism of this concept of Continental Drift, this contribution to the study of biogeography was significant in the way that it shed light on the importance of environmental and geographic similarities or differences as a result of climate and other pressures on the planet.

Alfred Russel Wallace studied the distribution of flora and fauna in the Amazon Basin and the Malay Archipelago in the mid-19th century. His research was essential to the further development of biogeography, and he was later nicknamed the "father of Biogeography". He and Philip Sclater saw biogeography as a source of support for the theory of evolution as they used Darwin's conclusion to explain how biogeography was similar to a record of species inheritance. Key findings, such as the sharp difference in fauna either side of the Wallace Line, and the sharp difference that existed between North andSouth America prior to their relatively recent faunal interchange, can only be understood in this light. Otherwise, the field of biogeography would be seen as a purely descriptive one.

The publication of *The Theory of Island Biogeography* by Robert MacArthur and E.O. Wilson in 1967 showed that the species richness of an area could be predicted in terms of such factors as habitat area, immigration rate and extinction rate.

This added to the long-standing interest in island biogeography. The application of island biogeography theory to habitat fragments spurred the development of the fields of conservation biology and landscape ecology.

Classic biogeography has been expanded by the development of molecular systematics, creating a new discipline known asphylogeography. This development allowed scientists to test theories about the origin and dispersal of populations, such asisland endemics. For example, while classic biogeographers were able to speculate about the origins of species in the Hawaiian Islands, phylogeography allows them to test theories of relatedness between these populations and putative source populations in Asia and North America.

Biogeography continues as a point of study for many life sciences and geography students worldwide, however it may be under different broader titles within institutions such as ecology or evolutionary biology.

MODERN APPLICATIONS OF BIOGEOGRAPHY

Biogeography now incorporates many different fields including but not limited to physical geography, geology, botany and plant biology, zoology, and general biology. A biogeographer's main focus is on what environmental factors and what the influence of man does to the distribution of the specific species of study. In terms of applications of biogeography as a science today, technological advances have allowed satellite imaging and processing of the Earth. Two main types of satellite imaging that are important within modern biogeography are Global Production Efficiency Model (GLO-PEM) and General Information Sensing (GIS). GLO-PEM uses satellite-imaging gives "repetitive, spatially contiguous, and time specific observations of vegetation." These observations are on a global scale. GIS can show certain processes on the earth's surface like whale locations, sea surface temperatures, and bathymetry. Current scientists also use coral reefs to delve into the history of biogeography through the fossilized reefs.

Paleobiogeography

Paleobiogeography goes one step further to include paleogeographic data and considerations of plate tectonics. Using molecular analyses and corroborated by fossils, it has been possible to demonstrate that perching birds evolved first in the region of Australia or the adjacent Antarctic (which at that time lay somewhat further north and had a temperate climate). From there, they spread to the other Gondwanan continents and Southeast Asia – the part of Laurasia then closest to their origin of dispersal – in the late Paleogene, before achieving a global distribution in the early Neogene. Not knowing the fact that at the time of dispersal, the Indian Ocean was much narrower than it is today, and that South America was closer to the Antarctic, one would be hard pressed to explain the presence of many "ancient" lineages of perching birds in Africa, as well as the mainly South American distribution of the suboscines.

Paleobiogeography also helps constrain hypotheses on the timing of biogeographic events such as vicariance and geodispersal, and provides unique information on the formation of regional biotas. For example, data from species-level phylogenetic and biogeographic studies tell us that the Amazonian fish fauna accumulated in increments over a period of tens of millions of years, principally by means of allopatric speciation, and in an arena extending over most of the area of tropical South America (Albert & Reis 2011). In other words, unlike some of the well-known insular faunas (Galapagos finches, Hawaiian drosophilid flies, African rift lake cichlids), the species-rich Amazonian ichthyofauna is not the result of recent adaptive radiations.

For freshwater organisms, landscapes are divided naturally into discrete drainage basins by watersheds, episodically isolated and reunited by erosional processes. In regions like the Amazon Basin with an exceptionally low (flat) topographic relief, the many waterways have had a highly reticulated history over geological time. In such a context, stream capture is an important factor affecting the evolution and distribution of freshwater organisms. Stream capture occurs when an upstream portion of one river drainage is diverted to the downstream portion of an adjacent basin. This can happen as a result of tectonic uplift (or subsidence), natural damming created by a landslide, or headward or lateral erosion of the watershed between adjacent basins.

Classification

Biogeography is a synthetic science, related to geography, biology, soil science, geology, climatology, ecology and evolution.

Some fundamental concepts in biogeography include:

- **Evolution** – change in genetic composition of a population
- **Extinction** – disappearance of a species
- **Dispersal** – movement of populations away from their point of origin, related to migration

Endemic areas

- **Geodispersal** – the erosion of barriers to biotic dispersal and gene flow, that permit range expansion and the merging of previously isolated biotas range and distribution
- **Vicariance** – the formation of barriers to biotic dispersal and gene flow, that tend to subdivide species and biotas, leading to speciation and extinction

Comparative biogeography

The study of comparative biogeography can follow two main lines of investigation:

- Systematic biogeography is the study of biotic area relationships, their distribution, and hierarchical classification;
- Evolutionary biogeography is the proposal of evolutionary mechanisms responsible for organismal distributions. Possible mechanisms include widespread taxa disrupted by continental break-up or individual episodes of long-distance movement.

8

Aquatic Animal Behaviour

Aquatic is the multidisciplinary study of aquatic systems, encompassing both freshwater and marine systems. Scientific investigations within this field often examine the human impact on and interaction with aquatic systems and range in scale from the molecular level of contaminants to the stresses on entire ecosystems.

Some of the major fields of study within aquatic sciences include: limnology (study of lakes, rivers, wetlands and groundwater); biogeochemistry; aquatic ecology; oceanography; marine biology; and hydrology.

MARINE ANIMAL CAPTIVITY END THE EXPLOITATION SOCIOLOGY

Marine animals have been held captive for hundreds of years and for a variety of reasons. As humans began to investigate the fascinating world below the ocean's surface, animal captivity became an everyday occurrence. Some animals have been captured as a means of research so scientists, as well as the public, can observe and learn more about them ("Marine Mammals in Captivity"). However, marine animals have also been exhibited simply for amusement and profit ("Do Marine Mammals Belong in Captivity in the 21st Century?"). Throughout history, humans have abused their relative power over marine creatures by capturing and detaining them. This cruel and unjust captivity commences with the act of capture and continues by diminishing marine animals' quality of life.

The crime of captivity begins as early as marine mammal capture. In the past, animal capture was a violent and traumatic process. Over the years, it has become progressively less malicious. However, the animals still suffer. Hunters herd the animals into shallow waters and proceed to entrap them in nets and slings ("Do Marine Mammals Belong in Captivity in the 21st Century?"). Captures can include high-speed chases intended to exhaust the animals, which makes them easier to catch. Some fisherman will actually ride the animals until they are completely worn out ("The Case Against Marine

Mammals in Captivity"). In Japan, fisherman are hired by captivity agencies to herd entire pods of dolphins so that the best and most promising mammals can be selected, while the remaining dolphins are slaughtered ("Global Ocean - Marine Mammal Anti-Captivity Officer"). In another instance, over 200 dolphins were driven into a fishing port, where they crashed into boats and each other. After becoming tangled in the chaos of nets, boats, and animals, many dolphins died of drowning ("Marine Animal Exhibits: Chlorinated Prisons"). Although the government works to regulate the individuals and organizations that capture marine animals, even the gentlest capture causes unforeseeable consequences.

Marine animals such as dolphins travel in groups and while it may seem like taking only one or two of the animals would cause no harm, it is extremely detrimental to the group as a whole. Whales, specifically orcas, are the largest animals held in captivity ("Marine Animal Exhibits: Chlorinated Prisons"). While orcas remain with their mothers for life in the wild, hunters often separate mother and child. Dolphins swim together in "pods," a family unit that consists of an adult dolphin and her offspring. These families are torn apart by captivity. Even if not all of the animals are captured, the free animals are left without a crucial member of their community. Some dolphins die simply from the stress of losing a family member or watching their companions being captured ("Marine Animal Exhibits: Chlorinated Prisons"). One study found that the mortality rate for bottlenose dolphins increased six-fold immediately after a capture ("The Case Against Marine Mammals in Captivity). The negative affects of animal capture are undeniable proof that holding marine animals captive is an unnatural and immoral act.

Even after the vicious and traumatic capture, marine animals continue to suffer in captivity. Although many trainers work to ensure that animal habitats are suitable for the animals they house, no man made structure can replace the natural habitat and ecosystem of the ocean. In addition, these artificial environments pose many risks to the animals they accommodate. First, the water of the tank can cause serious health risks. Many aquariums and marine parks pump water in directly from the ocean. However, this water is filtered and chlorine is added while micro and macro marine life is removed ("Killer Whales in Captivity"). This treatment creates harsh water full of chemicals, which can irritate the skin of marine mammals. Although the chemicals are used to purify the water, bacteria are still present and the animals' skin cannot tolerate the alien bacteria. Some dolphins go blind (French), while others animals suffer from skin diseases ("Marine Mammals in Captivity"). In other cases, orcas experience dorsal fin collapse. This occurs because the whales do not have the support of a large body of water, such as the ocean, and gravity pulls the tall appendage downward ("Marine Mammals in Captivity").

The issues are not limited to physical conditions. Being held in captivity actually affects the mental functions and capacity of marine creatures. Animals such as whales and dolphins utilize echolocation while living in the ocean. In this wide and varied body of water, these animals are constantly alert and exercising their brains. However, in a dull environment such as a small aquarium, these animals have no use for their highly evolved talent ("Marine Animal Exhibits: Chlorinated Prisons). Instead, they are forced to swim in circles without exercising the functions of their brain. For orcas, which are extremely sensitive to sound, the outside noises of water pumps and cheering crowds harm their hearing ("Killer Whales in Captivity"). Some studies show that dolphin brains shrink a frightening 42% while in captivity ("The Life of a Dolphin in Captivity"), and some dolphins have been driven insane by the constant reverberations of their own sonar waves that hit nothing but blank walls ("Marine Animal Exhibits: Chlorinated Prisons").

Not only are marine animals' brain functions halted, they also become socially and emotionally upset while in captivity. As mentioned earlier, dolphins and orcas are negatively affected by the separation of pods. Despite being social creatures that tend to have long term companionships, these mammals are separated from their families and isolated on their own when they are held captive. ("The Case Against Marine Mammals in Captivity"). The issues are furthered by trainers' treatment of the animals. For example, trainers will separate the acutely social dolphins when they misbehave, forcing them into isolation ("Marine Animal Exhibits: Chlorinated Prisons"). The torture continues in the feeding. In a technique called the "Pavlovian regime," trainers starve dolphins so that they will perform. Food is only administered as a reward to the dolphin for successful completion of tricks. Trainers effectively teach the dolphins that food is not a natural right of existence, but is instead only attained through submission and performance ("The Life of a Dolphin in Captivity"). It's also important to consider the food given to the animals - instead of live, freshly caught, natural marine organisms, captured marine animals are fed frozen fish and vitamin supplements ("Killer Whales in Captivity"). This unnatural diet hurts the metabolism of these creatures and hinders their instinctual predatory behavior.

Dolphins that live in captivity are forced to swim in circles in six-foot deep tanks that stretch twenty-four inches by twenty-for inches ("Marine Animal Exhibits: Chlorinated Prisons"). This lack of space literally suffocates the dolphins, who swim up to 100 miles each day in the wild. Finally, it's important to consider the behavior of freed marine animals. Dolphins and whales alike spend their days diving hundreds of meters, swimming hundreds of miles, and roaming freely about the ocean ("Marine Mammals in Captivity"). Unlike seals and sea lions, dolphins and whales rarely come up to the shore to perch and can stay underwater up to thirty minutes. The confinement of a tank forces a creature that previously spent 80-90% of its time underwater into a creature that is constantly above the water ("Marine Mammals in Captivity").

Some would argue that marine animal captivity has positive benefits for animals. For example, if a marine animal is held captive, humans are able to study and observe the animal, which in turn allows for a greater understanding of the species. This understanding allows humans to actually go out and assist the marine animals in the future. However, a marine animal held in captivity actually holds little educational value. These animals are forced to act differently than they do in the wild. Because they are confined to cages and tanks, they cannot roam and live as they would in the vast ocean. This means that when scientists observe an animal in a tank, he or she is not seeing the way the animal really acts, lives or behaves, but instead it's contrived adaptations to life in a tank ("The Case Against Marine Mammals in Captivity"). Others argue that holding marine animals in captivity saves them from the harsher environments in the wild and protects them from predators and pollution. However, this argument is incorrect. It is impossible for humans to judge what environment is too harsh for any particular animal. Marine animals have survived and evolved for thousands of years without human salvation or interaction and humans must allow this natural cycle to continue. In the wild, unhindered by human meddling, the evolutionary cycle will continue as it should and as is natural. Some species may become extinct or evolve into even more complex animals - this is not a negative progression but is instead the circle of life ("The Case Against Marine Mammals in Captivity"). While dolphins in their natural habitat can live to their forties and fifties, dolphins in aquariums and tanks often die before they reach twenty ("Marine Mammals in Captivity"). Over the years, nearly 4,000 sea lions, seals, and dolphins have died in captivity, and more than half of these deaths are human related. This includes things such as swallowing coins, dying of heat stroke, and swimming in contaminated water ("Marine Animal Exhibits: Chlorinated Prisons"). Although they claim to increase the longevity of marine animals' lives, those who capture marine animals end up hurting them and, in the long run, harming the natural flow of life.

It's certain that important information can be gained from marine animal captivity. However, holding wild animals hostage is immoral and unnecessary. Although humans are capable of capturing marine animals, this prevents them from existing in their natural habitat and only serves to hurt the species. To help stop captivity, it's important not to visit captive marine mammals in zoos or parks ("Marine Animal Exhibits: Chlorinated Prisons"). Also, instead of holding the animals hostage under the guise of saving them from even harsher natural environments, society should work to be environmentally conscious and preserve the animals' natural habitats, allowing them to live without the pollution of human waste. Holding marine animals in captivity is unequivocally wrong. No matter the claims of salvation and education, animal captivity is exploitation of animals.

Human and Aquatic Animal Communication

Ever since the beginning of animal and human existence, there has been a continuous flow of information between individuals and groups. The term communication comes from Latin word commu"nic re- "to share" and defines the process of giving information or of making emotions or ideas known to someone (MacMillan Dictionary n.d.).

Human communication has a very elaborated structure. A language is a complex communication system and it occurs when the speaker and the listener possess a representational process that is common and that ensures similar coding and decoding of signal meaning(Rendall et al. 2009) Even if animals might not have the ability to speak a language or utter various words, they have many other ways of expressing themselves.

In animal communication, scent is probably the most common sense used by animals in order to mark their territory, to warn off or even scare intruders, to show their readiness to mate or even to attract the prey. In general, they use strong smelling urine for these purposes but some animals possess scent glands that produce persistent odours. Skunks are widely known for their anal scent glands that spray an oily liquid towards their enemy. The foul mist does not do any serious damage to the victim but it scares him off and it could linger on his body for a few days. Snakes and lizards use their tongues in order to collect the scent particles in the air then they analyze the information inside their mouths. Many insects, such as bees, ants and moths use pheromones to communicate, mostly for attracting males. Another source of scent utilized by some animals (for example: the rabbit, the hippopotamus, the vicuna) to make their presence be felt on different territories is excrements. They try to mark their territory by dropping large heaps of feces so others would keep distance.

The second sense that animals rely on to a great extent is hearing. In order to survive and succeed, animals guide themselves by the acoustic signals that they produce and receive. There is an ample variety of sounds that could be produced by animals, depending on size and species. Numerous acoustic signals that are transmitted by animals are not distinguished or noticed by humans because of the ultra-low sounds that can't be perceived by human ear. The most vocal animals are birds that are well-known for their songs in different tonalities and cheerful chirping, especially in spring and summer. Whales, the largest of all mammals, are experts in the art of sound communication. They use various types of sounds, like whistles, clicks and pulsed calls for multiple purposes: echolocation, mate calling, and social interaction(Anon n.d.). Other animals that are often heard by humans are frogs and toads that have developed their vocal signals by using little bags of air as sound resonators which are situated on the sides of the head or in the throat. Elephants can produce a wide range of sounds, from rumbles to snorts and cries, going from very low frequencies to higher ones. When they are in danger or they are in a state of extreme excitement they use their trunks to deliver long resounding trumpeting. Their large ears allow them to have a great sensitivity to lower frequencies and also a better capacity in localizing sounds.

A third sense that helps animals communicate is sight. Visual signals at animals are diverse and they can come in many forms of display, such as mimicry, body posture, and facial expressions. For example, male peacocks and lyre birds display their exquisite colorful feathers and bodies in order to attract females for mating, visual communication being used in their advantage. Body language is frequently used by dogs, cats, monkeys and other animals. An example would be the moment when two dogs meet for the first time and they start setting their hierarchical position through body postures. If one holds its tail between the legs it means that the dog is found in a submissive position, allowing the other, with raised tail, to exercise power and dominance. Dogs also make use of their tails by wagging them in order to show happiness, excitement or pleasure. In opposition, cats waving tail is a signal of anger and irritation. The non-human primates, the monkeys, use a lot of facial expressions and body language to express their feelings. They flutter eyelids and raise eyebrows to denote pleasure, they hug to show friendship towards the other, and the males hit their chests in order to show superiority.

The art of mimicry is a fascinating modality of visual communication between animals. Mimicry occurs when an animal copies the actions, the appearance or the sounds of another animal or even the surrounding itself in order to avoid predators and enemies and to get them confused. For example, the peacock butterfly has large eye-spots on its wings to give the predator the false impression that it represents something bigger than it actually is so that would scare him off. Another example is the elephant hawk moth caterpillar who inflates its snake-like- head when being attacked to create the illusion of being a bigger and scarier animal. Other type of visual communication is warning coloration which can be found at caterpillars, various insects, frogs and snakes, the bright colors representing a warning for the potential predators. Their coloration pattern shows how dangerous or poisonous they can be and also if they have a dreadful taste. So, in general, animals learn how to avoid the danger in such cases.

Even though it might not be as important as other senses, touch is used by some animals in order to communicate their feelings towards others. As humans make certain gestures like shaking hands, kissing or hugging when meeting someone, monkeys also hug as a sign of affection. In addition to humans, they have a special and unique greeting which consists in placing the hand in the each other's mouth as a sign of trust and friendliness between them.

Cats are creatures that are avid after tactile communication with others of their kind or even with humans. They usually rub their bodies against each other, especially around the face area but also along their bodies, intertwining their tails. Some domestic and savage cats use each other as "cushions", this behavior being some form of social bonding and affection. They love grooming each other and curl up together. Owner of over 1000 tiny receptors, the crocodile is very sensitive to any kind of presence or movement. These receptors are present around their jaw line and they can easily detect vibrations in the water and the location of their prey even when a small insect stops by to take a drink.

Animal communication has always been fascinating to humans as we try to understand which are the similarities and differences between human language and animals' ways of expressing themselves.

Karl von Frisch, a professor of zoology at the University of Munich, conducted some research on bee communication during around thirty years and he revealed some important principles that helped us understand more about animal communication(Benveniste 1953) First of all, we found out that there are some similarities between bee communication and human language. They are able to offer and to receive messages that contain real data, they can describe the location and the distance of a certain object and they can communicate all these by using various somatic movements. Therefore, it was demonstrated that bees own some kind of "memory" which enables them to store all the data and then transmit it to others.

On the other hand, the differences between human language and bee communication are significant. Bees can only transmit the message through body language without being able to create sounds through vocal organs. Their communication can occur only during daylight, an element that permits visual perception while human communication is not limited by this factor.

Another important difference between the two 'languages' is the fact that, unlike humans, bees can't carry a dialogue therefore there is no exchange of linguistic information and no reply from the receiver of the message. Also, bees' messages can't be analyzed because they can only be seen as a general reference to a total content. But in human's speech, each utterance can be reduced to parts that could easily be combined according to some certain rules so there is a great variety in human language. So, we got to the conclusion that bee's communication is not a language but a signal code because there is an invariability of the message, a unilaterality of transmission, an impossibility of separating the elements of the message and also fixity of the topic.

Studies made on non-human primates showed that while humans acquire extra information about a speaker's intentions, wishes, and beliefs during a dialogue, apes and monkeys appear to have no such abilities because there is no theory of mind. Their way of communicating might appear a bit chaotic when it is compared to human language. For example, when they encounter dangerous situations, they create some alarm vocalizations which are not structured but short and noisy. These sounds are ideally produced to capture the listener's attention as quick as possible in order to warn it about the potential dangers and to make it react really fast.

One of the factors that led to animal abuse is neglect. Nowadays, many people want to keep a pet just because the animals are cute. People see those cute animals such as puppy, kitten and rabbit through mass media. The mass media include advertisements and some TV shows. After they saw those cute and adorable animals they want to keep it as pet, so they go to the pet shop to buy. They never think twice before they bring those animals back to their home. They never think that they don't have time to take care their pets. This will causes harms to their pets. In this 21st century most of the peoples and families are busy with their work, and they don't have time feed and play with their pets. When their pet is sick, they don't even notice and do not have time to bring their pet to the veterinarians. As a result, this will lead to animal cruelty.

Besides that, some of the TV shows and game shows also will lead to animal abuses. Some of the game shows will use animal as their selling point to attract viewers in order to boost up their viewership rating for the particular shows. Some of game shows even torture and hurt the animals in order to get good ratings for the show. For example, an American game show known as Fear Factor often uses animal to in their shows as a challenge to their game show's participants. In one of the episodes, the participants are required to use their mouth to move the snakes from one box to another box. This action will cause harm to the snakes and this is a very foolish and very unethical action done by human being.

Another factor that will lead to animal abuse or animal cruelty is the science research. In this century, science is very important for human being. Science is important to human beings in term of medications, technologies and others. In order to make humans life easier a lots of researches and experiments need to be carry out. Some of these experiments will involve animals. For example, in order to produce or invent new medicines animal will be the first one who test the drugs for the side effects. Besides, for education purpose in school or university, students are required to dissect animals such as rats and frogs in biology class in order to understand more about the system and facts about those animals. Animal abuse happens in this case indirectly.

In additions, stress and anger will is also one of the factors that will contribute to animal abuse. This phenomenon happens when someone got too stress from his work or studies and they have no way to release it, they will release it on their pets. Those people will torture their pets by not feeding their pet or even uses brutal strength such ass kick and hit their pets. Anger will also cause animal cruelty happens. When some weak people such as students got bullied and humiliate at school and they have no strength or power to protect and defend them self, they will get their revenge on animals by torturing them. They do this because they think those small animals like puppies and kittens are small and defenseless against them.

Another factor that will lead to animal abuse is the psychological factor. Some people out there have psychology and mental problem. They want the power to control people and dominate people, but they are not capable enough. So, they try to dominate the animals and control the animals by giving order and punishment as they want. Another type of psychology problem some people like to see animal being torture or they personally like to torture animal themselves for personal satisfaction. This is a kind of disease or mental disorder. People with this kind of mental disorder are mostly having trauma during their childhood time. These people are most likely being abuse by their parents or being bullied in school time. These people need medication to control their condition or else they will get worse. Another psychology problem that led to animal cruelty is the people who bully the animal to seek for public attention. These people are those who are commonly neglect by the society. They feel that they do not existence in the society and they feel that they are ignored. So they want to do something to let the public know about their existence. Some of these people will abuse and torture the animals to prove their existence in the society.

Some of the parents will buy pets to accompany their children. Parents nowadays are busy with their works and do not have time to play with their children. So they buy a pet for them to accompany them to play with them and make them happy. But they never thought that their children do not how to take care the animals. Sometimes is not they want to abuse the animals, but the will they treat the animals like feeding them too much will indirectly causes harm to the animals.

Another factor that will contribute to animal abuse is culture. Some of the culture will contribute to animal abuse. Those cultures are like a traditional festival for them and they must carry it out. For example Spain's bulls fight festival. This festival is a very cruel and unethical festival. This festival is actually a festival that kills a bull to entertain audiences. Besides that, some of the people will even kill those animals just for the skin of the animal. For example, human being kills those snakes, crocodiles and bulls just for their skin to make wallets and handbags.

Besides that, some human will also kill the animal for gaining personal profit. Shark fin is a very good example of animal cruelty. In order to get the shark fin, they will just cut off the fin of the shark and throw them back into the sea. The shark without the fin will hardly swim and eventually the sharks will die in the sea. Another example of this point is eating puppy or dogs. In China, they will slaughter the dogs and puppy in their shop and cook it for the customers. Eating dog is an unethical behavior to do.

In addition, animal training may also lead to animal abuse. To train an animal to obey to a command is not an easy task or more precise is a very hard task. All animals are different. Some of the people may eventually hurt the animal without noticing when they are training the animals. Some of the training needs punishment to let the animals know or remember the move the people teach them. For example, in a circus, animal such as tiger, lion, and elephant need training before they can perform in front of the audience, during the training the people may hurt them without noticing because animals cant express themselves well

HOMOSEXUAL BEHAVIOR IN AQUATICANIMALS

Homosexual behavior in animals is sexual behavior among non-human species that is interpreted as homosexual or bisexual. This may include sexual activity, courtship, affection, pair bonding, and parenting among same-sex animal pairs. Research indicates that various forms of this are found throughout the animal kingdom. As of 1999, about 500 species, ranging from primates to gut worms, have been documented engaging in same-sex behaviors. According to the organizers of the 2006 Against Nature? exhibit, it has been observed in 1,500 species.

According to Bruce Bagemihl, "the animal kingdom [does] it with much greater sexual diversity - including homosexual, bisexual and nonreproductive sex - than the scientific community and society at large have previously been willing to accept." Bagemihl adds, however, that this is "necessarily an account of human interpretations of these phenomena". Simon LeVay introduced the further caveat that "[a]lthough homosexual behavior is very common in the animal world, it seems to be very uncommon that individual animals have a long-lasting predisposition to engage in such behavior to the exclusion of heterosexual activities. Thus, a homosexual orientation, if one can speak of such thing in animals, seems to be a rarity." One species in which exclusive homosexual orientation occurs, however, is that of domesticated sheep (Ovis aries). "About 10% of rams (males) refuse to mate with ewes (females) but do readily mate with other rams."

The sexual behavior of non-human animals takes many different forms, even within the same species, though homosexual behavior is best known from social species. The motivations for and implications of these behaviors have yet to be fully understood, since most species have yet to be fully studied.

The observation of homosexual behavior in animals can be seen as both an argument for and against the acceptance of homosexuality in humans, and has been used especially against the claim that it is a peccatum contra naturam ("sin against nature"). For instance, homosexuality in animals was cited by the American Psychiatric Association and other groups in their amici curiae brief to the United States Supreme Court in Lawrence v. Texas, which ultimately struck down the sodomy laws of 14 states.

Applying The Term Homosexual to Animals

The term homosexual was coined by Karl-Maria Kertbeny in 1868 to describe same-sex sexual attraction and sexual behavior in humans. Its use in animal studies has been controversial for two main reasons: animal sexuality and motivating factors have been and remain poorly understood, and the term has strong cultural implications in western society that are irrelevant for species other than humans. Thus homosexual behavior has been given a number of terms over the years. When describing animals, the term homosexual is preferred over gay, lesbian, and other terms currently in use, as these are seen as even more bound to human homosexuality.

Animal preference and motivation is always inferred from behavior. In wild animals, researchers will as a rule not be able to map the entire life of an individual, and must infer from frequency of single observations of behavior. The correct usage of the term homosexual is that an animal exhibits homosexual behavior or even same-sex sexual behavior; however, this article conforms to the usage by modern research, applying the term homosexuality to all sexual behavior (copulation, genital stimulation, mating games and sexual display behavior) between animals of the same sex. In most instances, it is presumed that the homosexual behavior is but part of the animal's overall sexual behavioral repertoire, making the animal "bisexual" rather than "homosexual" as the terms are commonly understood in humans, but cases of homosexual preference and exclusive homosexual pairs are known.

Research

A majority of the research available concerning homosexual behavior in animals lacks specification between animals that exclusively exhibit same-sex tendencies and those that participate in heterosexual and homosexual mating activities interchangeably. This lack of distinction has led to differing opinions and conflicting interpretations of collected data amongst scientists and researchers. For instance, Bruce Bagemihl, author of the book Biological

Exuberence: Animal Homosexuality and Natural Diversity, emphasizes that there are no anatomical or endocrinological differences between exclusively homosexual and exclusively heterosexual animal pairs. However, if the definition of "homosexual behavior" is made to include animals that participate in both same-sex and opposite-sex mating activities, hormonal differences have been documented among key sex hormones, such as testosterone and estradiol, when compared to those who participate solely in heterosexual mating.

Many of the animals used in laboratory-based studies of homosexuality do not appear to spontaneously exhibit these tendencies often in the wild. Such behavior is often elicited and exaggerated by the researcher during experimentation through the destruction of a portion of brain tissue, or by exposing the animal to high levels of steroid hormones prenatally. Information gathered from these studies is limited when applied to spontaneously occurring same-sex behavior in animals outside of the laboratory.

The presence of same-sex sexual behavior was not "officially" observed on a large scale until recent times, possibly due to observer biascaused by social attitudes to same-sex sexual behavior, innocent confusion, or even from a fear of "being ridiculed by their colleagues".Georgetown University biologist Janet Mann states "Scientists who study the topic are often accused of trying to forward an agenda, and their work can come under greater scrutiny than that of their colleagues who study other topics." They also noted "Not every sexual act has a reproductive function... that's true of humans and non-humans." It appears to be widespread amongst social birds and mammals, particularly the sea mammals and the primates. The true extent of homosexuality in animals is not known. While studies have demonstrated homosexual behavior in a number of species, Petter Bøckman, the scientific advisor of the exhibition Against Nature? in2007, speculated that the true extent of the phenomenon may be much larger than was then recognized:

No species has been found in which homosexual behaviour has not been shown to exist, with the exception of species that never have sex at all, such as sea urchins and aphis. Moreover, a part of the animal kingdom is hermaphroditic, truly bisexual. For them, homosexuality is not an issue.

An example of overlooking homosexual behavior is noted by Bagemihl describing mating giraffes where nine out of ten pairings occur between males.

Every male that sniffed a female was reported as sex, while anal intercourse with orgasm between males was only "revolving around" dominance, competition or greetings.

Some researchers believe this behavior to have its origin in male social organization and social dominance, similar to the dominance traits shown in prison sexuality. Others, particularly Bagemihl, Joan Roughgarden, Thierry Lodé and Paul Vasey suggest the social function of sex (both homosexual and heterosexual) is not necessarily connected to dominance, but serves to strengthen alliances and social ties within a flock. Others have argued that social organization theory is inadequate because it cannot account for some homosexual behaviors, for example, penguin species where male individuals mate for life and refuse to pair with females when given the chance. While reports on many such mating scenarios are still only anecdotal, a growing body of scientific work confirms that permanent homosexuality occurs not only in species with permanent pair bonds, but also in non-monogamous species like sheep.

One report on sheep cited below states:

Approximately 8% of rams exhibit sexual preferences [that is, even when given a choice] for male partners (male-oriented rams) in contrast to most rams, which prefer female partners (female-oriented rams). We identified a cell group within the medial preoptic area/anterior hypothalamus of age-matched adult sheep that was significantly larger in adult rams than in ewes...

In fact, apparent homosexual individuals are known from all of the traditional domestic species, from sheep, cattle and horses to cats, dogs and budgerigars.

BASIS

Physiological Basis

A definite physiological explanation or reason for homosexual activity in animal species has not been agreed upon by researchers in the field. Numerous scholars are of the opinion that varying levels (either higher or lower) of the sex hormones in the animal, in addition to the size of the animal's gonads, play a direct role in the sexual behavior and preference exhibited by that animal. Others firmly argue no evidence to support these claims exists when comparing animals of a specific species exhibiting homosexual behavior exclusively and those that do not. Ultimately, empirical support from comprehensive endocrinological studies exist for both interpretations. Researchers found no evidence of differences in the measurements of the gonads, or the levels of the sex hormones of exclusively homosexual western gulls and ring-billed gulls. However, when analyzing these differences in bisexual rams, males were found to have lower levels of testosterone and estradiol in their blood, as well as smaller gonads than their heterosexual counterpart.

Additional studies pertaining to hormone involvement in homosexual behavior indicate that when administering treatments of testosterone and estradiol to female heterosexual animals, the elevated hormone levels increase the likelihood of homosexual behavior. Additionally, boosting the levels of sex hormones during an animal's pregnancy appears to increase the likelihood of it birthing a homosexual offspring.

Genetic Basis

Researchers found that disabling the fucose mutarotase (FucM) gene in laboratory mice - which influences the levels of estrogen to which the brain is exposed - caused the female mice to behave as if they were male as they grew up. "The mutant female mouse underwent a slightly altered developmental programme in the brain to resemble the male brain in terms of sexual preference" said Professor Chankyu Park of the Korea Advanced Institute of Science and Technology in Daejon, South Korea, who led the research. His most recent findings have been published in the BMC Genetics journal on July 7, 2010. Another study found that by manipulating a gene in fruit flies (Drosophila), homosexual behavior appeared to have been induced. However, in addition to homosexual behavior, several abnormal behaviors were also exhibited apparently due to this mutation.

Neurobiological Basis

In March 2011, research showed that serotonin is involved in the mechanism of sexual orientation of mice. A study conducted on fruit flies found that inhibiting the dopamine neurotransmitter inhibited lab-induced homosexual behavior.

SOME SELECTED SPECIES AND GROUPS

Birds

Black Swans

An estimated one-quarter of all black swans pairings are of homosexual males. They steal nests, or form temporary threesomes with females to obtain eggs, driving away the female after she lays the eggs. More of their cygnets survive to adulthood than those of different-sex pairs, possibly due to their superior ability to defend large portions of land. The same reasoning has been applied to male flamingo pairs raising chicks.

Gulls

Studies have shown that 10 to 15 percent of female western gulls in some populations in the wild exhibit homosexual behavior.

Ibises

Research has shown that the environmental pollutant methylmercury can increase the prevalence of homosexual behavior in male American white ibis. The study involved exposing chicks in varying dosages to the chemical and measuring the degree of homosexual behavior in adulthood. The results discovered was that as the dosage was increased the likelihood of homosexual behavior also increased. The endocrine blocking feature of mercury has been suggested as a possible cause of sexual disruption in other bird species.

Mallards

Two male mallards, Anas platyrhynchos Mallards form male-female pairs only until the female lays eggs, at which time the male leaves the female. Mallards have rates of male-male sexual activity that are unusually high for birds, in some cases, as high as 19% of all pairs in a population. Kees Moeliker of the Natural History Museum Rotterdam has observed one male mallard engage in homosexual necrophilia.

Penguins

Further information: Penguin § Behaviour, and Animal sexual behaviour § Penguins Penguins have been observed to engage in homosexual behaviour since at least as early as 1911. George Murray Levick, who documented this behaviour in Adélie penguins at Cape Adare, described it as "depraved". The report was considered too shocking for public release at the time, and was suppressed. The only copies that were made available privately to researchers were translated into Greek, to prevent this knowledge becoming more widely known. The report was unearthed only a century later, and published in Polar Record in June 2012.

In early February 2004 the New York Times reported that Roy and Silo, a male pair of chinstrap penguins in the Central Park Zoo in New York City had successfully hatched and fostered a female chick from a fertile egg they had been given to incubate. Other penguins in New York zoos have also been reported to have formed same-sex pairs.

In Odense Zoo in Denmark, a pair of male king penguins adopted an egg that had been abandoned by a female, proceeding to incubate it and raise the chick. Zoos in Japan and Germany have also documented homosexual male penguin couples. The couples have been shown to build nests together and use a stone as a substitute for an egg. Researchers at Rikkyo University in Tokyo found 20 homosexual pairs at 16 major aquariums and zoos in Japan.

The Bremerhaven Zoo in Germany attempted to encourage reproduction of endangered Humboldt penguins by importing females from Sweden and separating three male pairs, but this was unsuccessful. The zoo's director said that the relationships were "too strong" between the homosexual pairs. German gay groups protested at this attempt to break up the male-male pairs but the zoo's director was reported as saying "We don't know whether the three male pairs are really homosexual or whether they have just bonded because of a shortage of females... nobody here wants to forcibly separate homosexual couples."

A pair of male Magellanic penguins who had shared a burrow for six years at the San Francisco Zoo and raised a surrogate chick, split when the male of a pair in the next burrow died and the female sought a new mate. Buddy and Pedro, a pair of male African penguins, were separated by the Toronto Zoo to mate with female penguins. Buddy has since paired off with a female.

Suki and Chupchikoni are two female African penguins that pair bonded at the Ramat Gan Safari in Israel. Chupchikoni was assumed to be male until her blood was tested. In 2014 Jumbs and Hurricane, two Humboldt penguins at Wingham Wildlife Park became the center of international media attention as two male penguins who had pair bonded a number of years earlier and then successfully hatched and reared an egg given to them as surrogate parents after the mother abandoned it halfway through incubation.

VULTURES

In 1998 two male griffon vultures named Dashik and Yehuda, at the Jerusalem Biblical Zoo, engaged in "open and energetic sex" and built a nest. The keepers provided the couple with an artificial egg, which the two parents took turns incubating; and 45 days later, the zoo replaced the egg with a baby vulture. The two male vultures raised the chick together. A few years later, however, Yehuda became interested in a female vulture that was brought into the aviary. Dashik became depressed, and was eventually moved to the zoological research garden at Tel Aviv University where he too set up a nest with a female vulture.

Two male vultures at the Allwetter Zoo in Muenster built a nest together, although they were picked on and their nest materials were often stolen by other vultures. They were eventually separated to try to promote breeding by placing one of them with female vultures, despite the protests of German homosexual groups.

Pigeons

Both male and female pigeons sometimes exhibit homosexual behavior. In addition to sexual behavior, same-sex pigeon pairs will build nests, and hens will lay (infertile) eggs and attempt to incubate them.

Mammals

Amazon Dolphin

The Amazon river dolphin or boto has been reported to form up in bands of 3-5 individuals engaging in sexual activity. The groups usually comprise young males and sometimes one or two females. Sex is often performed in non-reproductive ways, using snout, flippers and genital rubbing, without regard to gender. In captivity, they have been observed to sometimes perform homosexual and heterosexual penetration of the blowhole, a hole homologous with the nostril of other mammals, making this the only known example of nasal sex in the animal kingdom. The males will sometimes also perform sex with males from the tucuxi species, a type of small porpoise.

Courtship, mounting, and full anal penetration between bulls has been noted to occur among American bison. The Mandan nation Okipa festival concludes with a ceremonial enactment of this behavior, to "ensure the return of the buffalo in the coming season". Also, mounting of one female by another (known as "bulling") is extremely common among cattle. The behaviour is hormone driven and synchronizes with the emergence of estrus (heat), particularly in the presence of a bull.

BONOBO AND OTHER APES

Bonobos, which have a matriarchal society, unusual among apes, are a fully bisexual species-both males and females engage in heterosexual and homosexual behavior, being noted for female-female homosexuality in particular. Roughly 60% of all bonobo sexual activity occurs between two or more females. While the homosexual bonding system in bonobos represents the highest frequency of homosexuality known in any species, homosexuality has been reported for all great apes (a group which includes humans), as well as a number of other primatespecies.

Dutch primatologist Frans de Waal on observing and filming bonobos noted that there were two reasons to believe sexual activity is the bonobo's answer to avoiding conflict. Anything that arouses the interest of more than one bonobo at a time, not just food, tends to result in sexual contact. If two bonobos approach a cardboard box thrown into their enclosure, they will briefly mount each other before playing with the box. Such situations lead to squabbles in most other species. But bonobos are quite tolerant, perhaps because they use sex to divert attention and to defuse tension.

Bonobo sex often occurs in aggressive contexts totally unrelated to food. A jealous male might chase another away from a female, after which the two males reunite and engage in scrotal rubbing. Or after a female hits a juvenile, the latter's mother may lunge at the aggressor, an action that is immediately followed by genital rubbing between the two adults.

Bottlenose Dolphins

Dolphins of several species engage in homosexual acts, though it is best studied in the bottlenose dolphins. Sexual encounters between females take the shape of "beak-genital propulsion", where one female inserts her beak in the genital opening of the other while swimming gently forward. Between males, homosexual behaviour includes rubbing of genitals against each other, which sometimes leads to the males swimming belly to belly, inserting the penis in the others genital slit and sometimes anus.

Janet Mann, Georgetown University professor of biology and psychology, argues that the strong personal behavior among male dolphin calves is about bond formation and benefits the species in an evolutionary context. She cites studies showing that these dolphins later in life as adults are in a sense bisexual, and the male bonds forged earlier in life work together for protection as well as locating females to reproduce with. Confrontations between flocks of bottlenose dolphins and the related species Atlantic spotted dolphin will sometimes lead to cross-species homosexual behaviour between the males rather than combat.

Elephants

African and Asian males will engage in same-sex bonding and mounting. Such encounters are often associated with affectionate interactions, such as kissing, trunk intertwining, and placing trunks in each other's mouths. Male elephants, who often live apart from the general herd, often form "companionships", consisting of an older individual and one or sometimes two younger males with sexual behavior being an important part of the social dynamic. Unlike heterosexual relations, which are always of a fleeting nature, the relationships between males may last for years. The encounters are analogous to heterosexual bouts, one male often extending his trunk along the other's back and pushing forward with his tusks to signify his intention to mount. Same-sex relations are common and frequent in both sexes, with Asiatic elephants in captivity devoting roughly 45% of sexual encounters to same-sex activity.

Giraffes

Male giraffes have been observed to engage in remarkably high frequencies of homosexual behavior. After aggressive "necking", it is common for two male giraffes to caress and court each other, leading up to mounting and climax. Such interactions between males have been found to be more frequent than heterosexual coupling. In one study, up to 94% of observed mounting incidents took place between two males. The proportion of same sex activities varied between 30 and 75%, and at any given time one in twenty males were engaged in non-combative necking behavior with another male. Only 1% of same-sex mounting incidents occurred between females.

Monkeys

Among monkeys, Lionel Tiger and Robin Fox conducted a study on how Depo-Provera contraceptives lead to decreased male attraction to females.

Japanese Macaque

With the Japanese macaque, also known as the "snow monkey", same-sex relations are frequent, though rates vary between troops. Females will form "consortships" characterized by affectionate social and sexual activities. In some troops up to one quarter of the females form such bonds, which vary in duration from a few days to a few weeks. Often, strong and lasting friendships result from such pairings. Males also have same-sex relations, typically with multiple partners of the same age. Affectionate and playful activities are associated with such relations.

Lions

Both male and female lions have been seen to interact homosexually. Male lions pair-bond for a number of days and initiate homosexual activity with affectionate nuzzling and caressing, leading to mounting and thrusting. About 8% of mountings have been observed to occur with other males. Pairings between females are held to be fairly common in captivity but have not been observed in the wild.

Polecat

European polecats Mustela putorius were found to engage homosexually with non-sibling animals. Exclusive homosexuality with mounting and anal penetration in this solitary species serves no apparent adaptive function.

Sheep

Ovis aries has attracted much attention due to the fact that around 8-10% of rams have an exclusive homosexual orientation. Furthermore, around 18-22% of rams are bisexual.

An October 2003 study by Dr. Charles E. Roselli et al. (Oregon Health and Science University) states that homosexuality in male sheep (found in 8% of rams) is associated with a region in the rams' brains which the authors call the "ovine Sexually Dimorphic Nucleus" (oSDN) which is half the size of the corresponding region in heterosexual male sheep.

Scientists found that, "The oSDN in rams that preferred females was significantly larger and contained more neurons than in male-oriented rams and ewes. In addition, the oSDN of the female-oriented rams expressed higher levels of aromatase, a substance that converts testosterone to estradiol, a form of estrogen which is believed to facilitate typical male sexual behaviors. Aromatase expression was no different between male-oriented rams and ewes."

"The dense cluster of neurons that comprise the oSDN express cytochrome P450 aromatase. Aromatase mRNA levels in the oSDN were significantly greater in female-oriented rams than in ewes, whereas male-oriented rams exhibited intermediate levels of expression." These results suggest that "... naturally occurring variations in sexual partner preferences may be related to differences in brain anatomy and its capacity for estrogen synthesis." As noted before, given the potential unagressiveness of the male population in question, the differing aromatase levels may also have been evidence of aggression levels, not sexuality. It should also be noted that the results of this study have not been confirmed by other studies.

The Merck Manual of Veterinary Medicine appears to consider homosexuality among sheep as a routine occurrence and an issue to be dealt with as a problem of animal husbandry.

Spotted Hyena

The family structure of the spotted hyena is matriarchal, and dominance relationships with strong sexual elements are routinely observed between related females. Due largely to the female spotted hyena's unique urogenital system, which looks more like a penis rather than a vagina, early naturalists thought hyenas were hermaphroditic males who commonly practiced homosexuality. Early writings such as Ovid's Metamorphoses and the Physiologus suggested that the hyena continually changed its sex and nature from male to female and back again. In Paedagogus, Clement of Alexandria noted that the hyena (along with the hare) was "quite obsessed with sexual intercourse". Many Europeans associated the hyena with sexual deformity, prostitution, deviant sexual behavior, and even witchcraft.

The reality behind the confusing reports is the sexually aggressive behavior between the females, including mounting between females. Research has shown that "in contrast to most other female mammals, female Crocuta are male-like in appearance, larger than males, and substantially more aggressive," and they have "been masculinized without being defeminized".

Study of this unique genitalia and aggressive behavior in the female hyena has led to the understanding that more aggressive females are better able to compete for resources, including food and mating partners. Research has shown that "elevated levels of testosterone in utero"contribute to extra aggressiveness; both males and females mount members of both the same and opposite sex, who in turn are possibly acting more submissive because of lower levels of testosterone in utero.

Others

Lizards

Whiptail lizard (Teiidae genus) females have the ability to reproduce through parthenogenesis and as such males are rare and sexual breeding non-standard. Females engage in sexual behavior to stimulate ovulation, with their behavior following their hormonal cycles; during low levels of estrogen, these (female) lizards engage in "masculine" sexual roles. Those animals with currently high estrogen levels assume "feminine" sexual roles. Lizards that perform the courtship ritual have greater fertility than those kept in isolation due to an increase in hormones triggered by the sexual behaviors. So, even though asexual whiptail lizards populations lack males, sexual stimuli still increase reproductive success. From an evolutionary standpoint, these females are passing their full genetic code to all of their offspring (rather than the 50% of genes that would be passed in sexual reproduction). Certain species of gecko also reproduce by parthenogenesis.

Insects and Arachnids

There is evidence of homosexual behavior in at least 110 species of insects and arachnids. Homosexual behavior in insects has been a matter of significant debate among scientists. Though various theories have been proposed, the leading explanation for why males, in some species up to 85%, engage in homosexuality is simply a case of mistaken identity. The cost of missing an opportunity to copulate with a female is greater than the risk of engaging in sexual activity with an insect or spider of the same sex.

Dragonflies

Male homosexuality has been inferred in several species of dragonflies (the order Odonata). The cloacal pinchers of male damselflies and dragonflies inflict characteristic head damage to females during sex. A survey of 11 species of damsel and dragonflies has revealed such mating damages in 20 to 80% of the males too, indicating a fairly high occurrence of sexual coupling between males.

Fruit Flies

Male Drosophila melanogaster flies bearing two copies of a mutant allele in the fruitless gene court and attempt to mate exclusively with other males. The genetic basis of animal homosexuality has been studied in the fly Drosophila melanogaster. Here, multiple genes have been identified that can cause homosexual courtship and mating. These genes are thought to control behavior through pheromones as well as altering the structure of the animal's brains. These studies have also investigated the influence of environment on the likelihood of flies displaying homosexual behavior.

9

Collective Animal Behavior

Collective animal behavior describes the coordinated behavior of large groups of similar animals as well as emergent properties of these groups. This can include the costs and benefits of group membership, the transfer of information across the group, the group decision-making process, and group locomotion and synchronization. Studying the principles of collective animal behavior has relevance to human engineering problems through the philosophy of biomimetics. For instance, determining the rules by which an individual animal navigates relative to its neighbors in a group can lead to advances in the deployment and control of groups of swimming or flying micro-robots such as UAVs.

THE COLLECTIVE UNCONSCIOUS

Carl Jung believed our psyche included 3 parts - the conscious, the personal unconscious and the collective unconscious.

He illustrated this by asking us to imagine our minds like tropical islands. The small part above the water is our conscious. Consciousness is the experience of the here and now. It is the smallest part of our psyche.

Jung considered the ego to be the centre of the field of consciousness and the ego is only concerned with conscious experience.

The part of the island below the water is described as the personal unconscious. The personal unconscious contains both the pre conscious and unconscious.

All that was once conscious to the person is stored in the subjective personal unconscious. Unlike Freud, Jung believed the personal unconscious did not include instincts. However, he did believe it contained suppressed material. Jung saw the unconscious as a reservoir of both repressed material and creativity

The Collective Unconscious is symbolised by the sea bed linking all the islands together. It is the unconscious part of our mind we inherit from our ancestors. It has also been described as our psychic inheritance. Within the collective unconscious lies the shared journey, a search for meaning and a reservoir of knowledge we are all born with. It influences all or our experiences and behaviour, especially our emotional ones.

Examples of experiences that show the effects of the Collective Unconscious are

- Love at first Sight
- De Ja Vu
- Immediate recognition of certain myths

Jung's archetypes are described as the contents of the collective unconscious mind. They act as an organising principle on the things we do or see. Archetypes are not biological drives as in Freud's theories but are more spiritual demands.

Examples of Archetypes

Mother Archetype - We come into this world ready to want a mother. The Mother Archetype is our in-built ability to recognise a certain relationship - that of mothering. Even when an archetype doesn't have a particular real person available, we tend to personify the archetype, that is, turn it into a mythological "story-book" character. This character symbolizes the archetype.

The Shadow

Sex & Life instincts are stored in the Shadow which comes from our pre human past. It is the dark side of our ego, and the evil we are capable of is stored in the shadow. The shadow is amoral, it is neither nor bad. It is like an animal it just does what it does in order to survive. It is the animal in us, one moment an animal can be nurturing to its litter, the next it is ferocious and savage in order to survive. Humans see the animal world as being brutal and savage, therefore, the shadow becomes something of a garbage can for the parts of ourselves we cannot admit to.

Persona

The persona is described as the mask you put on before you face the outside world

At its worst, it can be mistaken, even by ourselves, for our true nature. Sometimes we believe we really are what we pretend to be!

Anima and Animus is the part of our personal that play the role of male or female

The anima is the female aspect present in the collective unconscious of men and animus is the male aspect present in the collective unconscious of women..

According to Jung there are no fixed numbers of archetypes and many seem to overlap.

The self is the most important archetype of all. It is the unity of the personality.

THE DYNAMICS OF THE PSYCHE

Principle of Opposites - Every wish always immediately has an opposite. If I have a good thought then I cannot help but have a bad thought. In order to know good, you must know bad. According to Jung it is the opposition that creates the power in the psyche.

Principle of Equivalence - The energy created by the opposites is given to both sides equally (i.e. both thoughts good and bad). If you face your other choice, or wish then the energy that was in this choice will go towards self improvement. However, if you deny the other choice existed the energy will go towards creating a complex.

A Complex is a suppressed thought or feeling that clusters around some theme provided by an archetype. If you deny ever having a bad thought, you might put that thought in form offered by the Shadow. If you pretend all your life you are only good then all the times you do good that other side goes into complex around the Shadow. The complex will then develop a life of its own and it will haunt you.

Principle of Entropy - This is the tendency for opposites to come together or the even distribution of energy.

When we are young, the opposites will tend to be extreme, and so we tend to have lots of energy. For example, adolescents tend to exaggerate male-female differences, with boys trying hard to be macho and girls trying equally hard to be feminine. And so their sexual activity is invested with great amounts of energy! Plus, adolescents often swing from one extreme to another, being wild and crazy one minute and finding religion the next.

As we get older, most of us come to be more comfortable with our different facets. We are a bit less naively idealistic and recognize that we are all mixtures of good and bad. We are less threatened by the opposite sex within us and become more androgynous. Even physically, in old age, men and women become more alike. This process of rising above our opposites, of seeing both sides of who we are, is called transcendence.

The Self - the goal of life is to realize the self. The self is an archetype that represent the transcendence of all opposites. It is a more balanced position for the psyche

Understanding collective action in biological processes is a central challenge, essential for achieving progress in a variety of fields ranging from the organization and evolution of neural decision-making circuits, to the coordinated communication among cells, or animals, to the dynamics of information exchange among sophisticated organisms, and the emergence of complex societies. Consequently the study of collective behavior naturally spans scales, from how neural circuits control individual behavior, to the analogous issue of determining the structure and function of the communication network among organisms that gives rise to emergent group, and population-level, behavior. Insights and paradigms from the study of collective behavior help reveal unifying principles, as well as important differences, across scales of biological organization.

Animal communication is any behavior on the part of one animal that has an effect on the current or future behavior of another animal. Body language like some gesture and action is a common way that animal communicates with others. Combined with other body language, in a specific context, many gestures such as yawns, direction of vision, and so on all transmit meaning. As with human beings, who may smile or hug or stand a particular way for multiple reasons, many animals reuse gestures as well. Communication system varies among animals. Some may use gestures and some may use a particular sound to inform others some message. Therefore, in the following section, some examples of how different animal communicate with others will be provided.

Different Animal's Communication System

Bee

The communication system of bee is complex as it use gestural. Bees communicate with others by dancing. Bees dance when they have found food source. For example, a particular bee called forager bee has to locate the food source for the hive. When a forager bee finds food and returns to the hive, it does a special dance on the wall of the hive in order to inform the other bees of the location of the food so that they can go and get it. Depending on the orientation of the dance on the wall of hive, the bees know which direction from the sun to fly and how far to fly. The primary function of this dance is to recruit other forager bees to gather the food.

Bird

The communication system of bird is using both vocalizations and visualizations. Vocalization is a mainly communication method among birds. Therefore, we focus on vocalizations which are divided into two parts which are calls and songs. Bird calls are made up of single notes or short sequences of notes. Calls are used to signal a lot of activities For example, birds employ calls during flight to keep the flock together and to signal takeoff and landing. They also utilize them when they need to sound an alarm or to protect their territory. Birds use different types of calls depending on what threat they perceive. If there is a danger, the call and the reaction of the birds that hear the call will be different from usual. When they hear this call, birds either stop moving or get under cover so they are harder to catch.

Dolphins and Whales

The communication systems of dolphins and whales are complex and quiet similar to those of birds in their functions. For example, the bottle-nosed dolphin creates two kinds of vocalizations which are pure tones, comprised of whistles and squeaks and pulsed sounds which including clicks, barks, yelps and moans. Dolphins produce clicks to send out sound waves that they bounce off objects to identify them, a process called echolocation. Whistles are also used to send out alarms and distress calls. If a dolphin is injured, it can signal other dolphins to come and provide help by raising the injured one out of the water so it can breathe.

THE DIFFERENCES BETWEEN ANIMAL COMMUNICATION SYSTEM AND HUMAN LANGUAGE

The capacity to communicate is innate in both humans and other species of animals. However, there are significant differences between human language and animal communication systems.

For human, communication means that a sender provides information to a receiver through a signal to interpret the exchange. For animal,less about information, more about managing or manipulating behavior.

Human language possesses a number of significant qualities that as yet have not been found, as a group, as a part of any other animal communication system. For example, human speech sounds can be arranged in infinite sequences to create new meanings. No animal species has been identified that can combine the sounds and gestures of their system to add new meanings in an ongoing and productive way. Each and every day humans are constantly creating new words and topics of discussion.

In addition to this, the relationship between the sound of a word and its meaning is almost always arbitrary in human languages. This is evidenced by the fact that all human s are of the same species but that one species' individuals currently speak over 6000 languages. In animal communication systems, a specific sound corresponds with a specific meaning. For animals, not only is the ability to communicate innate, but also the sounds, gestures and postures are often biologically fixed according to each species.

Finally, humans do something quite significant with their languages that animals do not. We communicate in the abstract, about events in the future or the past, about things and people not in the immediate environment, and concepts and ideas that have no physical form. Although some animal communication may deal, with items not in the immediate environment, animal communication is limited to reacting to the environment in some way.

The Uniqueness of Human Language

People have made attempts to teach animals to speak human language. For example, one of the well known experiments with teaching animals language was Allen and Beatrice Gardner's work in the1960s and 1970s with Washoe, a chimpanzee. According ti the Gardner's, Washoe learned to make over 100 signs and could combine two of them together in meaningful ways, for example, 'you drink'.

However, animals can never communicate in the same way as human beings do. They may be able to produce sounds similar to those used in human language, but it cannot be equated with the ability to acquire the complex grammar of a human language.

According to Yule(2010), human language is unique and has the following five distinct features:

Displacement

Humans can refer to past and future time. This property of human language is called displacement. It allows language users to talk about things and events not present in the immediate environment. This cannot be done by animal. For example, If a cat stands at your feet calling "meow" when you comes home, you are likely to understand this message as relating to that immediate time and place. If you ask your cat where it has been and what it was up to, you will probably get the same response "meow". This shows that animal communication seems to be designed exclusively for this moment, here and now. It cannot effectively be used to relate events that are far removed in time and place.

For human, we can say "I went to China in the summer" but a cat can never say "I went to a friend's home last night". That is, human communication is not limited by time and space.

Arbitrariness

There is no "natural" connection between a linguistic form and its meaning. The connection is quite arbitrary. For example, there is no rational reason to explain why a four legged domestic animal making "woof" sound is called "dog" in English, "gou" in Chinese, and "hund" in German. The relationship between the sound and the meaning is completely random.

There are some words in language with sounds that seem to "echo" the sounds of objects or activities and hence seem to have a less arbitrary connection. For example, cat say "ngeong" in Indonesian, a rooster says "cock-a-doodle-doo" in English but "cocorico" in French or "kikiriki" in German.

Productivity

Humans are continually creating new expressions and novel utterances by manipulating their linguistic resources to describe new objects and situations. This property is described as productivity and essentially means that the potential number of utterances in any human language is infinite. However, the communicative systems of animals do not have this kind of flexibility. They are not able to invent new signals to communicate. Take honeybee as an example, the honeybee normally able to communicate the location of a nectar source to other bees, but it will fail to do so if the location is really "new". Each signal in the system is fixed as relating to a particular object or occasion.

Cultural transmission

The process whereby a language is passed on from one generation to the next is described as cultural transmission. It is clear human beings are born with the biological organs that can help them to produce speech sounds, they still have to acquire a language from the people around them. It is so different from animal as they are born with a set of fixed signals that are produced instinctively. For example, when a bird is isolated for the first two months without hearing other birds, it can still instinctively produce the same songs the other birds of the same species do. But human infants growing up in isolation without any human language input cannot speak a language instinctively.

Duality

Human language is organized at two levels or layers simultaneously. This property is called duality. In speech production, we have a physical level at which we can produce individual sounds, like n,b and i. As individual sounds, none of these discrete forms has any meaning. In a particular combination such as "bin", we have another level producing a meaning that is different from the meaning of the combination in "nib". Therefore, at one level we have distinct sounds and at another level, we have distinct meanings. Among other creatures, each communicative signal appears to be a single fixed form that cannot be broken down into separate parts. Although your dog may be able to produce "woof"("I am happy to see you), it does not seem to do so on the basis of a distinct level of production combining the separate elements of w+oo+f.

PROPOSED FUNCTIONS

Many functions of animal aggregations have been proposed. These proposed functions may be grouped into the four following categories: social and genetic, anti-predator, enhanced foraging, and increased locomotion efficiency.

Social Interaction

Support for the social and genetic function of aggregations, especially those formed by fish, can be seen in several aspects of their behavior. For instance, experiments have shown that individual fish removed from a school will have a higher respiratory rate than those found in the school. This effect has been partly attributed to stress, although hydrodynamic factors were considered more important in this particular study. The calming effect of being with conspecifics may thus provide a social motivation for remaining in an aggregation. Herring, for instance, will become very agitated if they are isolated from conspecifics. Fish schools have also been proposed to serve a reproductive function since they provide increased access to potential mates.

Protetion from Predators

Several anti-predator functions of animal aggregations have been proposed. One potential method by which fish schools or bird flocks may thwart predators is the 'predator confusion effect' proposed and demonstrated by Milinksi and Heller (1978). This theory is based on the idea that it becomes difficult for predators to pick out individual prey from groups because the many moving targets create a sensory overload of the predator's visual channel. Milinski and Heller's findings have been corroborated both in experiment and computer simulations.

A second potential anti-predator effect of animal aggregations is the "many eyes" hypothesis. This theory states that as the size of the group increases, the task of scanning the environment for predators can be spread out over many individuals. Not only does this mass collaboration presumably provide a higher level of vigilance, it could also allow more time for individual feeding.

A third hypothesis for an anti-predatory effect of animal aggregation is the "encounter dilution" effect. Hamilton, for instance, proposed that the aggregation of animals was due to a "selfish" avoidance of a predator and was thus a form of cover-seeking. Another formulation of the theory was given by Turner and Pitcher and was viewed as a combination of detection and attack probabilities. In the detection component of the theory, it was suggested that potential prey might benefit by living together since a predator is less likely to chance upon a single group than a scattered distribution. In the attack component, it was thought that an attacking predator is less likely to eat a particular animal when a greater number of individuals are present. In sum, an individual has an advantage if it is in the larger of two groups, assuming that the probability of detection and attack does not increase disproportionately with the size of the group.

Enhanced Foraging

A third proposed benefit of animal groups is that of enhanced foraging. This ability was demonstrated by Pitcher and others in their study of foraging behavior in shoaling cyprinids. In this study, the time it took for groups of minnows and goldfish to find a patch of food was quantified. The number of fishes in the groups was varied, and a statistically significant decrease in the amount of time necessary for larger groups to find food was established. Further support for an enhanced foraging capability of schools is seen in the structure of schools of predatory fish. Partridge and others analyzed the school structure of Atlantic bluefin tuna from aerial photographs and found that the school assumed a parabolic shape, a fact that was suggestive of cooperative hunting in this species (Partridge et al., 1983).

Increased Locomotion Efficiency

This theory states that groups of animals moving in a fluid environment may save energy when swimming or flying together, much in the way that bicyclists may draft one another in a peloton. Geese flying in a Vee formation are also thought to save energy by flying in the updraft of the wing tip vortex generated by the previous animal in the formation. Ducklings have also been shown to save energy by swimming in a line. Increased efficiencies in swimming in groups have also been proposed for schools of fish and Antarctic krill.

Group Structure

The structure of large animal groups has been difficult to study because of the large number of animals involved. The experimental approach is therefore often complemented by mathematical modeling of animal aggregations.

Experimental Approach

Experiments investigating the structure of animal aggregations seek to determine the 3D position of each animal within a volume at each point in time. It is important to know the internal structure of the group because that structure can be related to the proposed motivations for animal grouping.

This capability requires the use of multiple cameras trained on the same volume in space, a technique known as stereophotogrammetry. When hundreds or thousands of animals occupy the study volume, it becomes difficult to identify each one. In addition, animals may block one another in the camera views, a problem known as occlusion. Once the location of each animal at each point in time is known, various parameters describing the animal group can be extracted.

These parameters include

Density: The density of an animal aggregation is the number of animals divided by the volume (or area) occupied by the aggregation. Density may not be a constant throughout the group. For instance, starling flocks have been shown to maintain higher densities on the edges than in the middle of the flock, a feature that is presumably related to defense from predators. Polarity: The group polarity describes if the group animals are all pointing in the same direction or not. In order to determine this parameter, the average orientation of all animals in the group is determined. For each animal, the angular difference between its orientation and the group orientation is then found. The group polarity is then the average of these differences (Viscido 2004).

Nearest Neighbor Distance: The nearest neighbor distance (NND) describes the distance between the centroid of one animal (the focal animal) and the centroid of the animal nearest to the focal animal. This parameter can be found for each animal in an aggregation and then averaged. Care must be taken to account for the animals located at the edge of an animal aggregation. These animals have no neighbor in one direction.

Nearest Neighbor Position: In a polar coordinate system, the nearest neighbor position describes the angle and distance of the nearest neighbor to a focal animal.

Packing Fraction: Packing fraction is a parameter borrowed from physics to define the organization (or state i.e. solid, liquid, or gas) of 3D animal groups. It is an alternative measure to density. In this parameter, the aggregation is idealized as an ensemble of solid spheres, with each animal at the center of a sphere. The packing fraction is defined as the ratio of the total volume occupied by all individual spheres divided by the global volume of the aggregation (Cavagna 2008). Values range from zero to one, where a small packing fraction represents a dilute system like a gas. Cavagna found that the packing fraction for groups of starlings was 0.012.

Integrated Conditional Density: This parameter measures the density at various length scales and therefore describes the homogeneity of density throughout an animal group.

Pair Distribution Function: This parameter is usually used in physics to characterize the degree of spatial order in a system of particles. It also describes the density, but this measures describes the density at a distance away from a given point. Cavagna et al. found that flocks of starlings exhibited more structure than a gas but less than a liquid.

Modeling Approach

The simplest mathematical models of animal aggregations generally instruct the individual animals to follow three rules:

1. Move in the same direction as your neighbor
2. Remain close to your neighbors
3. Avoid collisions with your neighbors

An example of such a simulation is the Boids program created by Craig Reynolds in 1986. Another is the Self Propelled Particle model. Many current models use variations on these rules. For instance, many models implement these three rules through layered zones around each animal. In the zone of repulsion very close to the animal, the focal animal will seek to distance itself from its neighbors in order to avoid a collision. In the slightly further away zone of alignment, a focal animal will seek to align its direction of motion with its neighbors. In the outmost zone of attraction, which extends as far away from the focal animal as it is able to sense, the focal animal will seeks to move towards a neighbor. The shape of these zones will necessarily be affected by the sensory capabilities of the animal. For example, the visual field of a bird does not extend behind its body. Fish, on the other hand, rely on both vision and on hydrodynamic signals relayed through its lateral line. Antarctic krill rely on vision and on hydrodynamic signals relayed through its antennae.

Recent studies of starling flocks have shown, however, that each bird modifies its position relative to the six or seven animals directly surrounding it, no matter how close or how far away those animals are. Interactions between flocking starlings are thus based on a topological rule rather than a metric rule. It remains to be seen whether the same rule can be applied to other animals. Another recent study, based on an analysis of high speed camera footage of flocks above Rome and assuming minimal behavioural rules, has convincingly simulated a number of aspects of flock behaviour.

COLLECTIVE DECISION MAKING

Aggregations of animals are faced with decisions which they must make if they are to remain together. For a school of fish, an example of a typical decision might be which direction to swim when confronted by a predator.

Social insects such as ants and bees must collectively decide where to build a new nest. A herd of elephants must decide when and where to migrate. How are these decisions made? Do stronger or more experienced 'leaders' exert more influence than other group members, or does the group make a decision by consensus? The answer probably depends on the species. While the role of a leading matriarch in an elephant herd is well known, studies have shown that some animal species use a consensus approach in their collective decision-making process.

A recent investigation showed that small groups of fish used consensus decision-making when deciding which fish model to follow. The fish did this by a simple quorum rule such that individuals watched the decisions of others before making their own decisions. This technique generally resulted in the 'correct' decision but occasionally cascaded into the 'incorrect' decision. In addition, as the group size increased, the fish made more accurate decisions in following the more attractive fish model. Consensus decision-making, a form of collective intelligence, thus effectively uses information from multiple sources to generally reach the correct conclusion.

Collective Intelligence in Animals

Collective intelligence (CI) is shared or group intelligence that emerges from the collaboration, collective efforts, and competition of many individuals and appears in consensus decision making. The term appears in sociobiology, political science and in context of mass peer review and crowd sourcing applications. It may involve consensus, social capital and formalisms such as voting systems, social media and other means of quantifying mass activity. Collective IQ is a measure of collective intelligence, although it is often used interchangeably with the term collective intelligence. Collective intelligence has also been attributed to bacteria and animals.

It can be understood as an emergent property from the synergies among: 1) data-information-knowledge; 2) software-hardware; and 3) experts (those with new insights as well as recognized authorities) that continually learns from feedback to produce just-in-time knowledge for better decisions than these three elements acting alone. Or more narrowly as an emergent property between people and ways of processing information. This notion of collective intelligence is referred to as "symbiotic intelligence" by Norman Lee Johnson. The concept is used in sociology, business, computer science and mass communications: it also appears in science fiction. Pierre Lévy defines collective intelligence as, "It is a form of universally distributed intelligence, constantly enhanced, coordinated in real time, and resulting in the effective mobilization of skills. I'll add the following indispensable characteristic to this definition: The basis and goal of collective intelligence is mutual recognition and enrichment of individuals rather than the cult of fetishized or hypostatized communities." According to researchers Pierre Lévy and Derrick de Kerckhove, it refers to capacity of networked ICTs (Information communication technologies) to enhance the collective pool of social knowledge by simultaneously expanding the extent of human interactions.

Collective intelligence strongly contributes to the shift of knowledge and power from the individual to the collective. According to Eric S. Raymond (1998) and JC Herz (2005), open source intelligence will eventually generate superior outcomes to knowledge generated by proprietary software developed within corporations (Flew 2008). Media theorist Henry Jenkins sees collective intelligence as an 'alternative source of media power', related to convergence culture. He draws attention to education and the way people are learning to participate in knowledge cultures outside formal learning settings. Henry Jenkins criticizes schools which promote 'autonomous problem solvers and self-contained learners' while remaining hostile to learning through the means of collective intelligence. Both Pierre Lévy (2007) and Henry Jenkins(2008) support the claim that collective intelligence is important for democratization, as it is interlinked with knowledge-based culture and sustained by collective idea sharing, and thus contributes to a better understanding of diverse society.

Similar to the g factor (g) for general individual intelligence, a new scientific understanding of collective intelligence aims to extract a general collective intelligence factor c factor for groups indicating a group's ability to perform a wide range of tasks. Definition, operationalization and statistical methods are derived from g. Similarly as g is highly interrelated with the concept of IQ, this measurement of collective intelligence can be interpreted as intelligence quotient for groups (Group-IQ) even though the score is not a quotient per se. Causes for c and predictive validity are investigated as well.

he concept (although not so named) originated in 1785 with the Marquis de Condorcet, whose "jury theorem" states that if each member of a voting group is more likely than not to make a correct decision, the probability that the highest vote of the group is the correct decision increases with the number of members of the group (see Condorcet's jury theorem). Many theorists have interpreted Aristotle's statement in the Politics that "a feast to which many contribute is better than a dinner provided out of a single purse" to mean that just as many may bring different dishes to the table, so in a deliberation many may contribute different pieces of information to generate a better decision. Recent scholarship, however, suggests that this was probably not what Aristotle meant but is a modern interpretation based on what we now know about team intelligence. A precursor of the concept is found in entomologist William Morton Wheeler's observation that seemingly independent individuals can cooperate so closely as to become indistinguishable from a single organism (1911). Wheeler saw this collaborative process at work in ants that acted like the cells of a single beast he called a superorganism.

In 1912 Émile Durkheim identified society as the sole source of human logical thought. He argued in "The Elementary Forms of Religious Life" that society constitutes a higher intelligence because it transcends the individual over space and time. Other antecedents are Vladimir Vernadsky's concept of "noosphere" and H.G. Wells's concept of "world brain" (see also the term "global brain"). Peter Russell, Elisabet Sahtouris, and Barbara Marx Hubbard (originator of the term "conscious evolution") are inspired by the visions of a noosphere - a transcendent, rapidly evolving collective intelligence - an informational cortex of the planet. The notion has more recently been examined by the philosopher Pierre Lévy. In a 1962 research report, Douglas Engelbart linked collective intelligence to organizational effectiveness, and predicted that pro-actively 'augmenting human intellect' would yield a multiplier effect in group problem solving: "Three people working together in this augmented mode [would] seem to be more than three times as effective in solving a complex problem as is one augmented person working alone". In 1994, he coined the term 'collective IQ' as a measure of collective intelligence, to focus attention on the opportunity to significantly raise collective IQ in business and society.

The idea of collective intelligence also forms the framework for contemporary democratic theories often referred to as epistemic democracy. Epistemic democratic theories refer to the capacity of the populace, either through deliberation or aggregation of knowledge, to track the truth and relies on mechanisms to synthesize and apply collective intelligence.

Dimensions

Howard Bloom has discussed mass behavior - collective behavior from the level of quarks to the level of bacterial, plant, animal, and human societies. He stresses the biological adaptations that have turned most of this earth's living beings into components of what he calls "a learning machine". In 1986 Bloom combined the concepts of apoptosis, parallel distributed processing, group selection, and the superorganism to produce a theory of how collective intelligence works. Later he showed how the collective intelligences of competing bacterial colonies and human societies can be explained in terms of computer-generated "complex adaptive systems" and the "genetic algorithms", concepts pioneered by John Holland.

Bloom traced the evolution of collective intelligence to our bacterial ancestors 1 billion years ago and demonstrated how a multi-species intelligence has worked since the beginning of life. Ant societies exhibit more intelligence, in terms of technology, than any other animal except for humans and co-operate in keeping livestock, for example aphids for "milking". Leaf cutters care for fungi and carry leaves to feed the fungi.

David Skrbina cites the concept of a 'group mind' as being derived from Plato's concept of panpsychism (that mind or consciousness is omnipresent and exists in all matter). He develops the concept of a 'group mind' as articulated by Thomas Hobbes in "Leviathan" and Fechner's arguments for a collective consciousness of mankind. He cites Durkheim as the most notable advocate of a "collective consciousness" and Teilhard de Chardin as a thinker who has developed the philosophical implications of the group mind.

Tom Atlee focuses primarily on humans and on work to upgrade what Howard Bloom calls "the group IQ". Atlee feels that collective intelligence can be encouraged "to overcome 'groupthink' and individual cognitive bias in order to allow a collective to cooperate on one process - while achieving enhanced intellectual performance." George Pór defined the collective intelligence phenomenon as "the capacity of human communities to evolve towards higher order complexity and harmony, through such innovation mechanisms as differentiation and integration, competition and collaboration." Atlee and Pór state that "collective intelligence also involves achieving a single focus of attention and standard of metrics which provide an appropriate threshold of action". Their approach is rooted in scientific community metaphor.

The term group intelligence is sometimes used interchangeably with the term collective intelligence. Anita Woolley presents Collective intelligence as a measure of group intelligence and group creativity. The idea is that a measure of collective intelligence covers a broad range of features of the group, mainly group composition and group interaction. The features of composition that lead to increased levels of collective intelligence in groups include criteria such as higher numbers of women in the group as well as increased diversity of the group.

Atlee and Pór suggest that the field of collective intelligence should primarily be seen as a human enterprise in which mind-sets, a willingness to share and an openness to the value of distributed intelligence for the common good are paramount, though group theory and artificial intelligence have something to offer. Individuals who respect collective intelligence are confident of their own abilities and recognize that the whole is indeed greater than the sum of any individual parts. Maximizing collective intelligence relies on the ability of an organization to accept and develop "The Golden Suggestion", which is any potentially useful input from any member.

Groupthink often hampers collective intelligence by limiting input to a select few individuals or filtering potential Golden Suggestions without fully developing them to implementation.

Robert David Steele Vivas in The New Craft of Intelligence portrayed all citizens as "intelligence minutemen," drawing only on legal and ethical sources of information, able to create a "public intelligence" that keeps public officials and corporate managers honest, turning the concept of "national intelligence" (previously concerned about spies and secrecy) on its head.

According to Don Tapscott and Anthony D. Williams, collective intelligence is mass collaboration. In order for this concept to happen, four principles need to exist;

Openness

Sharing ideas and intellectual property: though these resources provide the edge over competitors more benefits accrue from allowing others to share ideas and gain significant improvement and scrutiny through collaboration.

Peering

Horizontal organization as with the 'opening up' of the Linux program where users are free to modify and develop it provided that they make it available for others. Peering succeeds because it encourages self-organization - a style of production that works more effectively than hierarchical management for certain tasks.

Sharing

Companies have started to share some ideas while maintaining some degree of control over others, like potential and critical patent rights. Limiting all intellectual property shuts out opportunities, while sharing some expands markets and brings out products faster.

Acting Globally

The advancement in communication technology has prompted the rise of global companies at low overhead costs. The internet is widespread, therefore a globally integrated company has no geographical boundaries and may access new markets, ideas and technology.

COLLECTIVE INTELLIGENCE FACTOR C

A new scientific understanding of collective intelligence defines it as a group's general ability to perform a wide range of tasks. Definition, operationalization and statistical methods are similar to the psychometric approach of general individual intelligence. Hereby, an individual's performance on a given set of cognitive tasks is used to measure general cognitive ability indicated by the general intelligence factor g extracted via factor analysis. In the same vein as g serves to display between-individual performance differences on cognitive tasks, collective intelligence research aims to find a parallel intelligence factor for groups 'c factor' (also called 'collective intelligence factor' (CI)) displaying between-group differences on task performance. The collective intelligence score then is used to predict how this same group will perform on any other similar task in the future. Yet tasks, hereby, refer to mental or intellectual tasks performed by small groups even though the concept is hoped to be transferrable to other performances and any groups or crowds reaching from families to companies and even whole cities. Since individuals' g factor scores are highly correlated with full-scale IQ scores, which are in turn regarded as good estimates of g, this measurement of collective intelligence can also be seen as an intelligence indicator or quotient respectively for a group (Group-IQ) parallel to an individual's intelligence quotient (IQ) even though the score is not a quotient per se.

Mathematically, c and g are both variables summarizing positive correlations among different tasks supposing that performance on one task is comparable with performance on other similar tasks. c thus is a source of variance among groups and can only be considered as a group's standing on the c factor compared to other groups in a given relevant population. The concept is in contrast to competing hypotheses including other correlational structures to explain group intelligence, such as a composition out of several equally important but independent factors as found in individual personality research.

Besides, this scientific idea also aims to explore the causes affecting collective intelligence, such as group size, collaboration tools or group members' interpersonal skills. The MIT Center for Collective Intelligence, for instance, announced the detection of The Genome of Collective Intelligence as one of its main goals aiming to develop a taxonomy of organizational building blocks, or genes, that can be combined and recombined to harness the intelligence of crowds.

Causes

Individual intelligence is shown to be genetically and environmentally influenced. Analogously, collective intelligence research aims to explore reasons why certain groups perform more intelligent than other groups given that c is just moderately correlated with the intelligence of individual group members. According to Woolley et al.'s results, neither team cohesion nor motivation or satisfaction correlated with c. However, they claim that three factors were found as significant correlates: the variance in the number of speaking turns, group members' average social sensitivity and the proportion of females. All three had similar predictive power for c, but only social sensitivity was statistically significant (b=0.33, P=0.05).

The number speaking turns indicates that "groups where a few people dominated the conversation were less collectively intelligent than those with a more equal distribution of conversational turn-taking". Hence, providing multiple team members the chance to speak up made a group more intelligent.

Group members' social sensitivity was measured via the Reading the Mind in the Eyes Test (RME) and correlated. 26 with c. Hereby, participants are asked to detect thinking or feeling expressed in other peoples' eyes presented on pictures and assessed in a multiple choice format. The test aims to measure peoples' theory of mind (ToM), also called 'mentalizing' or 'mind reading', which refers to the ability to attribute mental states, such as beliefs, desires or intents, to other people and in how far people understand that others have beliefs, desires, intentions or perspectives different from their own ones. RME is a ToM test for adults that shows sufficient test-retest reliability and constantly differentiates control groups from individuals with functional autism or Asperger Syndrome. It is one of the most widely accepted and well-validated tests for ToM within adults. ToM can be regarded as an associated subset of skills and abilities within the broader concept of emotional intelligence.

The proportion of females as a predictor of c was largely mediated by social sensitivity (Sobel z = 1.93, P= 0.03) which is in vein with previous research showing that women score higher on social sensitivity tests. While a mediation, statistically speaking, clarifies the mechanism underlying the relationship between a dependent and an independent variable, Wolley agreed in an interview with the Harvard Business Review that these findings are saying that groups of women are smarter than groups of men. However, she relativizes this stating that the actual important thing is the high social sensitivity of group members.

It is theorized that the collective intelligence factor c is an emergent property resulting from bottom-up as well as top-down processes.Hereby, bottom-up processes cover aggregated group-member characteristics. Top-down processes cover group structures and norms that influence a group's way of collaborating and coordinating.

Processes

Top-Down Processes

Top-down processes cover group interaction, such as structures, processes, and norms. An example of such top-down processes is conversational turn-taking. Research further suggest that collectively intelligent groups communicate more in general as well as more equally; same applies for participation and is shown for face-to-face as well as online groups communicating only via writing.

Bottom-Up Processes

Bottom-up processes include group composition, namely the characteristics of group members which are aggregated to the team level encompassing. An example of such bottom-up processes is the average social sensitivity or the average and maximum intelligence scores of group members. Furthermore, collective intelligence was found to be related to a group's cognitive diversity including thinking styles and perspectives. Groups that are moderately diverse in cognitive style have higher collective intelligence than those who are very similar in cognitive style or very different. Consequently, groups where members are too similar to each other lack the variety of perspectives and skills needed to perform well. On the other hand, groups whose members are too different seem to have difficulties to communicate and coordinate effectively.

SERIAL VS PARALLEL PROCESSES

For most of human history, collective intelligence was confined to small tribal groups in which opinions were aggregated through real-time parallel interactions among members. In modern times, mass communication, mass media, and networking technologies have enabled collective intelligence to span massive groups, distributed across continents and time-zones. To accommodate this shift in scale, collective intelligence in large-scale groups been dominated by serialized polling processes such as aggregating up-votes, likes, and ratings over time. While modern systems benefit from larger group size, the serialized process has been found to introduce substantial noise that distorts the collective output of the group. In one significant study of serialized collective intelligence, it was found that the first vote contributed to a serialized voting system can distort the final result by 34%.

To address the problems of serialized aggregation of input among large-scale groups, recent advancements collective intelligence have worked to replace serialized votes, polls, and markets, with parallel systems such as "human swarms" modeled after synchronous swarms in nature. Based on natural process of Swarm Intelligence, these artificial swarms of networked humans enable participants to work together in parallel to answer questions and make predictions as an emergent collective intelligence. In one high profile example, a human swarm challenge by CBS Interactive to predict the Kentucky Derby. The swarm correctly predicted the first four horses, in order, defying 542-1 odds and turning a $20 bet into $10,800.

Evidnce

Woolley, Chabris, Pentland, Hashmi, & Malone (2010), the originators of this scientific understanding of collective intelligence, found a single statistical factor for collective intelligence in their research across 192 groups with people randomly recruited from the public. In Woolley et al.'s two initial studies, groups worked together on different tasks from the McGrath Task Circumplex, a well-established taxonomy of group tasks. Tasks were chosen from all four quadrants of the circumplex and included visual puzzles, brainstorming, making collective moral judgments, and negotiating over limited resources. The results in these tasks were taken to conduct a factor analysis. Both studies showed support for a general collective intelligence factor c underlying differences in group performance with an initial eigenvalue accounting for 43% (44% in study 2) of the variance, whereas the next factor accounted for only 18% (20%). That fits the range normally found in research regarding a general individual intelligence factor g typically accounting for 40% to 50% percent of between-individual performance differences on cognitive tests. Afterwards, a more complex criterion task was absolved by each group measuring whether the extracted c factor had predictive power for performance outside the original task batteries. Criterion tasks were playing checkers (draughts) against a standardized computer in the first and a complex architectural design task in the second study. In a regression analysis using both individual intelligence of group members and c to predict performance on the criterion tasks, c had a significant effect, but average and maximum individual intelligence had not. While average ($r=0.15$, $P=0.04$) and maximum intelligence ($r=0.19$, $P=0.008$) of individual group members were moderately correlated with c, c was still a much better predictor of the criterion tasks. According to Woolley et al., this supports the existence of a collective intelligence factor c, because it demonstrates an effect over and beyond group members' individual intelligence and thus that c is more than just the aggregation of the individual IQs or the influence of the group member with the highest IQ.

Engel et al. (2014) replicated Woolley et al.'s findings applying an accelerated battery of tasks with a first factor in the factor analysis explaining 49% of the between-group variance in performance with the following factors explaining less than half of this amount. Moreover, they found a similar result for groups working together online communicating only via text and confirmed the role of female proportion and social sensitivity in causing collective intelligence in both cases. Similarly to Wolley et al., they also measured social sensitivity with the RME which is actually meant to measure people's ability to detect mental states in other peoples' eyes. The online collaborating participants, however, did neither know nor see each other at all. The authors conclude that scores on the RME must be related to a broader set of abilities of social reasoning than only drawing inferences from other people's eye expressions.

A collective intelligence factor c in the sense of Woolley et al. was further found in groups of MBA students working together over the course of a semester, in online gaming groups as well as in groups from different cultures and groups in different contexts in terms of short-term versus long-term groups. None of these investigations considered team members' individual intelligence scores as control variables.

Note as well that the field of collective intelligence research is quite young and published empirical evidence is relatively rare yet. However, various proposals and working papers are in progress or already completed but (supposedly) still in a scholarly peer reviewing publication process.

Predictive Validity

Next to predicting a group's performance on more complex criterion tasks as shown in the original experiments, the collective intelligence factor c was also found to predict group performance in diverse tasks in MBA classes lasting over several months. Thereby, highly collectively intelligent groups earned significantly higher scores on their group assignments although their members did not do any better on other individually performed assignments. Moreover, highly collective intelligent teams improved performance over time suggesting that more collectively intelligent teams learn better. This is another potential parallel to individual intelligence where more intelligent people are found to acquire new material quicker.

Individual intelligence can be used to predict plenty of life outcomes from school attainment and career success to health outcomesand even mortality. Whether collective intelligence is able to predict other outcomes besides group performance on mental tasks has still to be investigated.

POTENTIAL CONNECTIONS TO INDIVIDUAL INTELLIGENCE

Gladwell (2008) showed that the relationship between individual IQ and success works only to a certain point and that additional IQ points over an estimate of IQ 120 do not translate into real life advantages. If a similar border exists for Group-IQ or if advantages are linear and infinite, has still to be explored. Similarly, demand for further research on possible connections of individual and collective intelligence exists within plenty of other potentially transferable logics of individual intelligence, such as, for instance, the development over time or the question of improving intelligence. Whereas it is controversial whether human intelligence can be enhanced via training, a group's collective intelligence potentially offers simpler opportunities for improvement by exchanging team members or implementing structures and technologies. Moreover, social sensitivity was found to be, at least temporarily, improvable by reading literary fiction as well as watching drama movies. In how far such training ultimately improves collective intelligence through social sensitivity remains an open question.

There are further more advanced concepts and factor models attempting to explain individual cognitive ability including the categorization of intelligence in fluid and crystallized intelligence or the hierarchical model of intelligence differences. Further supplementing explanations and conceptualizations for the factor structure of the Genomes' of collective intelligence besides a general c factor', though, are missing yet.

Controversies

Other scholars explain team performance by aggregating team members' general intelligence to the team level instead of building an own overall collective intelligence measure. Devine and Philips (2001) showed in a meta-analysis that mean cognitive ability predicts team performance in laboratory settings (.37) as well as field settings (.14) - note that this is only a small effect. Suggesting a strong dependence on the relevant tasks, other scholars showed that tasks requiring a high degree of communication and cooperation are found to be most influenced by the team member with the lowest cognitive ability. Tasks in which selecting the best team member is the most successful strategy, are shown to be most influenced by the member with the highest cognitive ability.

Since Woolley et al.'s results do not show any influence of group satisfaction, group cohesiveness, or motivation, they, at least implicitly, challenge these concepts regarding the importance for group performance in general and thus contrast meta-analytically proven evidence concerning the positive effects of group cohesion, motivation and satisfaction on group performance.

Noteworthy is also that the involved researchers among the confirming findings widely overlap with each other and with the authors participating in the original first study around Anita Woolley.

ALTERNATIVE MATHEMATICAL TECHNIQUES

Computational Collective Intelligence

In 2001, Tadeusz (Tad) Szuba from the AGH University in Poland proposed a formal model for the phenomenon of collective intelligence. It is assumed to be an unconscious, random, parallel, and distributed computational process, run in mathematical logic by the social structure.

In this model, beings and information are modeled as abstract information molecules carrying expressions of mathematical logic. They are quasi-randomly displacing due to their interaction with their environments with their intended displacements. Their interaction in abstract computational space creates multi-thread inference process which we perceive as collective intelligence. Thus, a non-Turing model of computation is used. This theory allows simple formal definition of collective intelligence as the property of social structure and seems to be working well for a wide spectrum of beings, from bacterial colonies up to human social structures. Collective intelligence considered as a specific computational process is providing a straightforward explanation of several social phenomena. For this model of collective intelligence, the formal definition of IQS (IQ Social) was proposed and was defined as "the probability function over the time and domain of N-element inferences which are reflecting inference activity of the social structure". While IQS seems to be computationally hard, modeling of social structure in terms of a computational process as described above gives a chance for approximation. Prospective applications are optimization of companies through the maximization of their IQS, and the analysis of drug resistance against collective intelligence of bacterial colonies.

Collective Intelligence Quotient

One measure sometimes applied, especially by more artificial intelligence focused theorists, is a "collective intelligence quotient" (or "cooperation quotient") - which can be normalized from the "individual" intelligence quotient (IQ) - thus making it possible to determine the marginal intelligence added by each new individual participating in the collective action, thus using metrics to avoid the hazards of group thinkand stupidity.

Applications

Elicitation of point estimates - Here, we try to get an estimate (in a single value) of something. For example, estimating the weight of an object, or the release date of a product or probability of success of a project etc. as are seen in prediction markets like Intrade, HSX or Inkling Markets and also in several implementations of crowd sourced estimation of a numeric outcome.

Essentially, we try to get the average value of the estimates provided by the members in the crowd.

Opinion Aggregation - In this situation, we gather opinions from the crowd regarding some idea, issue or product. For example, trying to get a rating (on some scale) of a product sold online (such as Amazon's star rating system). Here, the emphasis is to collect and simply aggregate the ratings provided by customers/users.

Idea Collection - In these problems, someone solicits ideas for projects, designs or solutions from the crowd. For example, ideas on solving a data science problem (as in Kaggle) or getting a good design for a t-shirt (as in Threadless) or in getting answers to simple problems that only humans can do well (as in Amazon's Mechanical Turk). Here, the objective is to gather the ideas and devise some selection criteria to choose the best ideas.

Cognition

Market Judgment

Because of the Internet's ability to rapidly convey large amounts of information throughout the world, the use of collective intelligence to predict stock prices and stock price direction has become increasingly viable. Websites aggregate stock market information that is as current as possible so professional or amateur stock analysts can publish their viewpoints, enabling amateur investors to submit their financial opinions and create an aggregate opinion. The opinion of all investor can be weighed equally so that a pivotal premise of the effective application of collective intelligence can be applied: the masses, including a broad spectrum of stock market expertise, can be utilized to more accurately predict the behavior of financial markets.

Collective intelligence underpins the efficient-market hypothesis of Eugene Fama - although the term collective intelligence is not used explicitly in his paper. Fama cites research conducted by Michael Jensen in which 89 out of 115 selected funds underperformed relative to the index during the period from 1955 to 1964. But after removing the loading charge (up-front fee) only 72 underperformed while after removing brokerage costs only 58 underperformed. On the basis of such evidence index funds became popular investment vehicles using the collective intelligence of the market, rather than the judgement of professional fund managers, as an investment strategy.

Cooperation

Networks of Trust

In 2012, the Global Futures Collective Intelligence System (GFIS) was created by The Millennium Project, which epitomizes collective intelligence as the synergistic intersection among data/information/knowledge, software/hardware, and expertise/insights that has a recursive learning process for better decision-making than the individual players alone.

New media are often associated with the promotion and enhancement of collective intelligence. The ability of new media to easily store and retrieve information, predominantly through databases and the Internet, allows for it to be shared without difficulty. Thus, through interaction with new media, knowledge easily passes between sources (Flew 2008) resulting in a form of collective intelligence. The use of interactive new media, particularly the internet, promotes online interaction and this distribution of knowledge between users.

Francis Heylighen, Valentin Turchin, and Gottfried Mayer-Kress are among those who view collective intelligence through the lens of computer science and cybernetics. In their view, the Internet enables collective intelligence at the widest, planetary scale, thus facilitating the emergence of a global brain.

The developer of the World Wide Web, Tim Berners-Lee, aimed to promote sharing and publishing of information globally. Later his employer opened up the technology for free use. In the early '90s, the Internet's potential was still untapped, until the mid-1990s when 'critical mass', as termed by the head of the Advanced Research Project Agency (ARPA), Dr. J.C.R. Licklider, demanded more accessibility and utility. The driving force of this Internet-based collective intelligence is the digitization of information and communication. Henry Jenkins, a key theorist of new media and media convergence draws on the theory that collective intelligence can be attributed to media convergence and participatory culture (Flew 2008). He criticizes contemporary education for failing to incorporate online trends of collective problem solving into the classroom, stating "whereas a collective intelligence community encourages ownership of work as a group, schools grade individuals". Jenkins argues that interaction within a knowledge community builds vital skills for young people, and teamwork through collective intelligence communities contribute to the development of such skills. Collective intelligence is not merely a quantitative contribution of information from all cultures, it is also qualitative.

Lévy and de Kerckhove consider CI from a mass communications perspective, focusing on the ability of networked information and communication technologies to enhance the community knowledge pool. They suggest that these communications tools enable humans to interact and to share and collaborate with both ease and speed (Flew 2008). With the development of the Internet and its widespread use, the opportunity to contribute to knowledge-building communities, such as Wikipedia, is greater than ever before. These computer networks give participating users the opportunity to store and to retrieve knowledge through the collective access to these databases and allow them to "harness the hive" Researchers at the MIT Center for Collective Intelligence research and explore collective intelligence of groups of people and computers.

In this context collective intelligence is often confused with shared knowledge. The former is the sum total of information held individually by members of a community while the latter is information that is believed to be true and known by all members of the community. Collective intelligence as represented by Web 2.0 has less user engagement than collaborative intelligence. An art project using Web 2.0 platforms is "Shared Galaxy", an experiment developed by an anonymous artist to create a collective identity that shows up as one person on several platforms like MySpace, Facebook, YouTube and Second Life. The password is written in the profiles and the accounts named "Shared Galaxy" are open to be used by anyone. In this way many take part in being one. Another art project using collective intelligence to produce artistic work is Curatron, where a large group of artists together decides on a smaller group that they think would make a good collaborative group. The process is used based on an algorithm computing the collective preferences In creating what he calls 'CI-Art', Nova Scotia based artist Mathew Aldred follows Pierry Lévy's definition of collective intelligence. Aldred's CI-Art event in March 2016 involved over four hundred people from the community of Oxford, Nova Scotia, and internationally. Later work developed by Aldred used the UNU swarm intelligence system to create digital drawings and paintings. The Oxford Riverside Gallery (Nova Scotia) held a public CI-Art event in May 2016, which connected with online participants internationally.

In social book marking (also called collaborative tagging), users assign tags to resources shared with other users, which gives rise to a type of information organisation that emerges from this crowdsourcing process. The resulting information structure can be seen as reflecting the collective knowledge (or collective intelligence) of a community of users and is commonly called a "Folksonomy", and the process can be captured by models of collaborative tagging.

Recent research using data from the social bookmarking website Delicious, has shown that collaborative tagging systems exhibit a form of complex systems (or self-organizing) dynamics. Although there is no central controlled vocabulary to constrain the actions of individual users, the distributions of tags that describe different resources has been shown to converge over time to a stable power law distributions. Once such stable distributions form, examining the correlations between different tags can be used to construct simple folksonomy graphs, which can be efficiently partitioned to obtained a form of community or shared vocabularies. Such vocabularies can be seen as a form of collective intelligence, emerging from the decentralised actions of a community of users. The Wall-it Project is also an example of social bookmarking.

P2P BUSINESS

Research performed by Tapscott and Williams has provided a few examples of the benefits of collective intelligence to business:

Talent Utilization

At the rate technology is changing, no firm can fully keep up in the innovations needed to compete. Instead, smart firms are drawing on the power of mass collaboration to involve participation of the people they could not employ. This also helps generate continual interest in the firm in the form of those drawn to new idea creation as well as investment opportunities.

Demand Creation

Firms can create a new market for complementary goods by engaging in open source community. Firms also are able to expand into new fields that they previously would not have been able to without the addition of resources and collaboration from the community. This creates, as mentioned before, a new market for complementary goods for the products in said new fields.

Costs Reduction

Mass collaboration can help to reduce costs dramatically. Firms can release a specific software or product to be evaluated or debugged by online communities. The results will be more personal, robust and error-free products created in a short amount of time and costs. New ideas can also be generated and explored by collaboration of online communities creating opportunities for free R&D outside the confines of the company.

OPEN SOURCE SOFTWARE

Cultural theorist and online community developer, John Banks considered the contribution of online fan communities in the creation of the Trainz product. He argued that its commercial success was fundamentally dependent upon "the formation and growth of an active and vibrant online fan community that would both actively promote the product and create content- extensions and additions to the game software".

The increase in user created content and interactivity gives rise to issues of control over the game itself and ownership of the player-created content. This gives rise to fundamental legal issues, highlighted by Lessig and Bray and Konsynski, such as intellectual property and property ownership rights.

Bibliography

- Albert, J. S., & R. E. Reis (2011). *Historical Biogeography of Neotropical Freshwater Fishes*. University of California Press, Berkeley. 424 pp.
- Alcock, J. (2009). *Animal Behavior: An Evolutionary Approach* (9th edition). Sinauer Associates Inc. Sunderland, MA.
- Behavioral Ecology. International Society for Behavioral Ecology. Retrieved 15 April 2011.
- Carson, R. (2002). *Silent Spring*. H7oughton Mifflin Company. p. 348. ISBN 0-618-24906-0.
- Cook, R. E. (1977). "Raymond Lindeman and the trophic-dynamic concept in ecology". *Science* 198 (4312): 22–26.Bibcode:1977Sci...198...22C.
- Danchin, É., Girladeau, L.-A. and Cézilly, F. (2008). *Behavioural Ecology: An Evolutionary Perspective on Behaviour*. Oxford University Press, Oxford.
- Ghilarov, A. M. (1995). "Vernadsky's biosphere concept: an historical perspective". *The Quarterly Review of Biology*70 (2): 193–203.
- Hamner, W. M. (1985). "The importance of ethology for investigations of marine zooplankton". *Bulletin of Marine Science* 37 (2): 414–424.
- Ives, A. R.; Cardinale, B. J.; Snyder, W. E. (2004). "A synthesis of subdisciplines: Predator–prey interactions, and biodiversity and ecosystem functioning". *Ecology Letters*8 (1): 102–116.
- Karban, R. (2008). "Plant behaviour and communication". *Ecology Letters* 11 (7): 727–739.
- Krebs, J. R.; Davies, N. B. (1993). *An Introduction to Behavioural Ecology*. Wiley-Blackwell. p. 432. ISBN 978-0-632-03546-5.
- MacArthur, Robert H. (1972). *Geographic Ecology*. New York: Harper & Row.
- Odum, E. P. (1968). "Energy flow in ecosystems: A historical review". *American Zoologist* 8 (1): 11–18.
- Palamar, C. R. (2008). "The justice of ecological restoration: Environmental history, health, ecology, and justice in the United States". *Human Ecology Review* 15(1): 82–94.

- Sakurai, K. (1985). "An attelabid weevil (*Euops splendida*) cultivates fungi". *Journal of Ethology* 3 (2): 151–156.
- Strassmann, J. E.; Zhu, Y.; Queller, D. C. (2000). "Altruism and social cheating in the social amoeba*Dictyostelium discoideum*". *Nature* 408 (6815): 965–967.
- Tinb1ergen, N. (1963). "On aims and methods of ethology" (PDF). Zeitschrift für Tierpsychologie 20 (4): 410–433.
- Webb, J. K.; Pike, D. A.; Shine, R. (2010). "Olfactory recognition of predators by nocturnal lizards: safety outweighs thermal benefits". Behavioural Ecology 21 (1): 72–77.
- Wilson, E. O. (2000). Sociobiology: The New Synthesis (25th anniversary ed.). President and Fellows of Harvard College. ISBN 978-0-674-00089-6.

Index

www.ingramcontent.com/pod-product-compliance
Ingram Content Group UK Ltd.
Pitfield, Milton Keynes, MK11 3LW, UK
UKHW021328190726
13845UKWH00004B/74

9 789349 964921